T0177992

Springer Tracts in Civil Engineering

Springer Tracts in Civil Engineering (STCE) publishes the latest developments in Civil Engineering - quickly, informally and in top quality. The series scope includes monographs, professional books, graduate textbooks and edited volumes, as well as outstanding PhD theses. Its goal is to cover all the main branches of civil engineering, both theoretical and applied, including:

- Construction and Structural Mechanics
- Building Materials
- Concrete, Steel and Timber Structures
- Geotechnical Engineering
- Earthquake Engineering
- Coastal Engineering; Ocean and Offshore Engineering
- Hydraulics, Hydrology and Water Resources Engineering
- Environmental Engineering and Sustainability
- Structural Health and Monitoring
- Surveying and Geographical Information Systems
- Heating, Ventilation and Air Conditioning (HVAC)
- Transportation and Traffic
- Risk Analysis
- Safety and Security

Indexed by Scopus

To submit a proposal or request further information, please contact:
Pierpaolo Riva at Pierpaolo.Riva@springer.com (Europe and Americas) Mengchu Huang at mengchu.huang@springer.com (China)

More information about this series at http://www.springer.com/series/15088

Marco Guerrieri · Raffaele Mauro

A Concise Introduction to Traffic Engineering

Theoretical Fundamentals and Case Studies

Marco Guerrieri
DICAM
University of Trento
Trento, Italy

Raffaele Mauro
DICAM
University of Trento
Trento, Italy

ISSN 2366-259X ISSN 2366-2603 (electronic)
Springer Tracts in Civil Engineering
ISBN 978-3-030-60725-8 ISBN 978-3-030-60723-4 (eBook)
https://doi.org/10.1007/978-3-030-60723-4

This Springer imprint is published by the registered company Springer Nature Switzerland AG
The registered company address is: Gewerbestrasse 11, 6330 Cham, Switzerland

Preface

This book contains a selection of fundamentals topics of traffic engineering useful for highways facilities design and control.

The treatment is basic, but it does not neglect to illustrate the most recent and crucial theoretical aspects which are at the root of several Highway Engineering applications, like, for instance, the essential aspects of highways traffic stream reliability calculation and automated highway systems control.

In order for these topics to be better understood, varied illustrative examples of applications are provided in great detail. We have actually sought to use an intuitive and discursive, rather than formal and abstract style throughout the book.

Shortly, the contents are as follows:

- Chapter 1 provides the definition of macroscopic traffic variables and the deduction of the fundamental traffic flow relationship;
- Chapter 2 offers an intuitive definition of transport demand, capacity and flow. The main macroscopic flow models, levels of service and traffic hysteresis phenomena are also dealt with;
- Chapter 3 gives an introduction to the flow continuity equation, dynamic traffic flow models and kinematic and shock waves. Since these topics are usually difficult for beginners, the mathematical treatment is highly detailed;
- Chapter 4 describes some of the main microscopic models (linear and nonlinear). These dynamic traffic models involve studying local instability, asymptotic instability and flow breakdown. Moreover, a genealogy of the main traffic models (macroscopic, mesoscopic and microscopic) is reviewed in a synthetic way;
- Chapter 5 deals with the fundamentals of random and traffic processes and provides a very accurate description of probability models for arrival, speed, headway and vehicular density processes. Also the main counting, headway and speed probability distributions are shown;
- Chapter 6 presents an advanced method for estimating flow reliability on highways and describes the current systems of highway traffic management and control. In this chapter, a rigorous capacity definition is offered. The basic

characteristics of the Automated Highway System are also introduced. Finally, the HSM method to estimate the annual crash frequency expected on highways and the COPERT method to calculate the polluting emissions are both illustrated;

- Chapter 7 presents the gap acceptance theory, in that it is of great applicative interest in studying road intersections. Therefore, some models for estimating the critical gap and the follow-up time for at-grade unsignalized intersections and roundabouts are described;
- Chapter 8 deals with the models for studying queues in road transport systems under a unitary approach. They are the probabilistic models for stationary state, the deterministic solutions in congestion, and the heuristic solutions for stationary and non-stationary states. The treatment of these models is unusual but can be directly applied to practical cases;
- Chapter 9 deals with unsignalized intersections and methods for determining the measures of effectiveness (MOE): waiting times and delays. To this regard, the TRL method for estimating capacity, queues, delays at three-arm intersections is exemplified;
- Chapter 10 covers signalized intersections. The HCM model for the calculation and functional analysis of such intersections is illustrated in detail.

Andrea Pompigna, Ph.D. in Transportation Engineering, is the author of Chap. 8. We are deeply grateful to Andrea Pompigna also for his invaluable help in discussing critically the topics with us, as well as revising the whole book in a systematic way.

Finally, we wish to thank Autostrade del Brennero S.p.A. (Trento) for supporting our recent research on traffic flow theory and control at DICAM, University of Trento.

Palermo, Italy — Marco Guerrieri
Palma Campania (Naples), Italy — Raffaele Mauro
July 2020

Contents

List of Figures

List of Tables

Chapter 1
Macroscopic Variables and Fundamental Relationships of Traffic Flow Theory

Abstract This chapter introduces some fundamental variables of traffic flow theory. These variables allow to develop traffic models for a specific class: macroscopic or in-mean-field models (see Chaps. 2 and 3). Though simple, models are highly interesting for practical applications (e.g. level of service and traffic flow control, see Chaps. 2 and 6).

1.1 Interrupted and Uninterrupted Traffic Flows

Traffic conditions on highways facilities can be of two types [1]:

- *uninterrupted flow*: the traffic flow can be perturbed only by internal factors, never by causes external to the traffic stream;
- *interrupted flow*: the traffic flow can be perturbed or stopped by causes external to the traffic stream (e.g. opposing vehicle flows, right-of-way systems at intersections, traffic lights etc.).

The uninterrupted flow always occurs on freeways, on road sections sufficiently far from on- and off-ramps (or nearer, if the on-ramp is controlled with traffic light systems (i.e. ramp metering, see Chap. 6) [2].

The uninterrupted flow can also occur on two-way roads but only on those sections which are not affected by perturbations due to manoeuvres at at-grade intersections (acceleration, deceleration, vehicle standstill, etc.) [3]. Conversely, the interrupted flow mainly occurs in urban road systems. More in detail, it concerns all types of intersections and their arms.

Finally, it is worth observing that both terms, 'interrupted flow' and 'uninterrupted flow', refer to the highway-type facilities. Traffic quality, on the other hand, is described with such terms as free, conditioned, congestioned, stable, unstable etc.

M. Guerrieri and R. Mauro, *A Concise Introduction to Traffic Engineering*,
Springer Tracts in Civil Engineering,
https://doi.org/10.1007/978-3-030-60723-4_1

1.2 Inference of Macroscopic Flow Variables

Consider a traffic stream on a lane (or a carriageway). For a generic vehicle "i" of this stream the following variables are defined (see Fig. 1.1a, b):

- length: l_i;
- position, referred to a vehicle point (e.g. bumper): abscissa x_i, with respect to the x-axis coinciding with the lane axis, or the carriageway axis. The axis is positively oriented in the direction of motion;
- instantaneous speed: $v_i = \frac{dx_i}{dt}$;
- longitudinal acceleration: $a_i = \frac{dv_i}{dt} = \frac{d^2x_i}{dt^2}$.

x_i, v_i and a_i are the kinematic variables as defined in physics.

When a vehicle follows another in the traffic stream, in the direction of motion, we denote the *leader vehicle* with "i" and the *follower vehicle* with "i + 1". The positions of "i" and "i + 1", referred to their rear bumpers, are x_i and x_{i+1} respectively.

Between the two consecutive vehicles there is a space headway h_{si}.

h_{si} is equal to the rear/front distance between vehicles, or *space gap* g_{si} plus the length of the vehicle follower l_{i+1} (Fig. 1.1a). Thus, the result is:

$$h_{si} = x_i - x_{i+1} = g_{si} + l_{i+1}.$$

Also a *time headway* h_{ti} can be measured between consecutive vehicles. h_{ti} is the time between the passing of the rear bumper of vehicles "i" on a cross road section of abscissa x and that of follower vehicles i + 1 on the same location x. Therefore (see Fig. 1.1b), h_{ti} is the sum of the *time gap* g_{ti} and the occupation time ρ_i.

g_{ti} is the time between the rear of the lead vehicle and the front of the following vehicle.

ρ_i is the time for the entire vehicle to pass x.

Thus, the result is:

- time headway: $h_{ti} = g_{ti} + \rho_i$

If the vehicle speed is constant (parallel trajectories, Fig. 1.1b), it follows:

$$\frac{h_{si}}{h_{ti}} = \frac{g_{si}}{g_{ti}} = \frac{l_i}{\rho_i} \tag{1.1}$$

Should these analyses be made on vehicles in a front-to-front position, they will produce the same results as above.

Define now three domains in t − x plane (Fig. 1.2):

- *time domain* D_t: to measure the kinematic variables in a time interval ΔT on a short segment Δx long. Δx straddles a generic road cross section with abscissa x (location x). For instance, the D_t domain is used for measuring instantaneous speeds by means of traffic detectors (piezoelectric sensors, inductive loops, magnetic sensors, radar, etc.) on freeway sections;

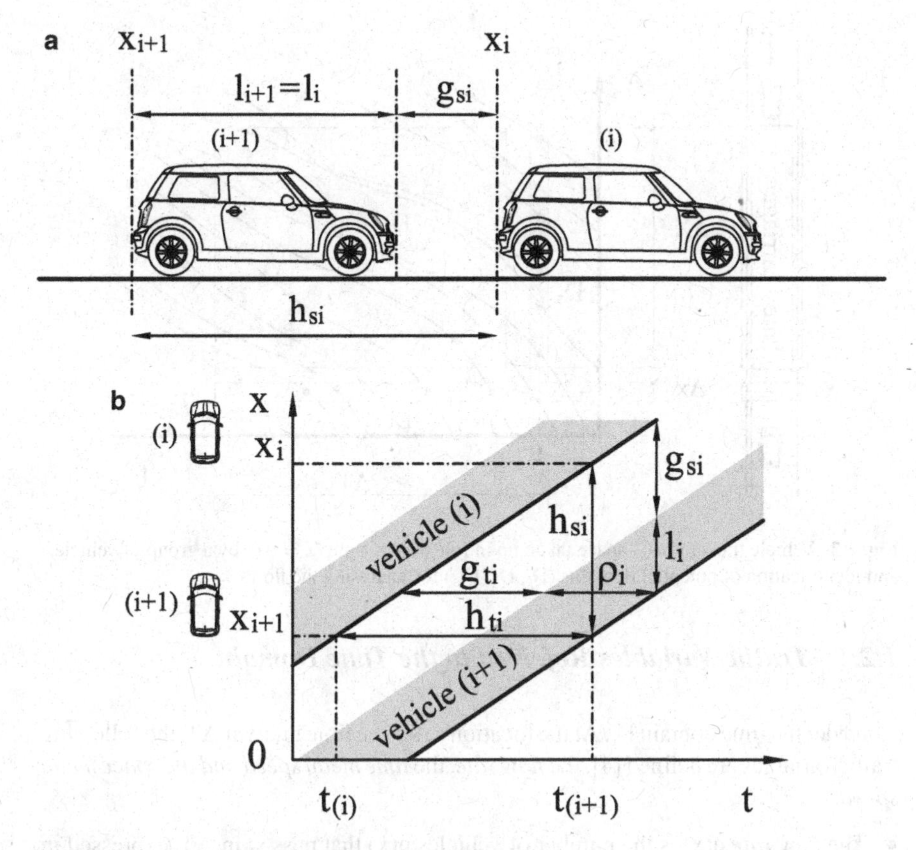

Fig. 1.1 Space and time headways between vehicles

- *space domain* D_s: to count the vehicles on a segment ΔX long, over an elementary time interval Δt, straddling a generic instant t. The observation and calculation is based e.g. by aerial photographs taken from above the road section under study;
- *time/space domain* $D_{t,s}$: to observe the vehicle trajectories $x_i = x_i(t)$ i = 1,2, … n along a segment ΔX long over a time interval ΔT. Once the trajectories in $D_{t,s}$ are known, vehicle speeds $v_i = v_i(t)$ and accelerations $a_i = a_i(t)$ can be computed at each x and t. The $D_{t,s}$ domain is obtained e.g. by video recordings.

From the definitions of D_t, D_s and $D_{t,s}$, it follows that $D_{t,s}$ is a D_t domain if ΔX is small, $\Delta X = \Delta x$ (elementary segment); $D_{t,s}$ is a D_s domain if ΔT is small, $\Delta T = \Delta t$ (elementary time interval). The small width of Δx and Δt is set depending on the problems under study.

When all vehicles have the same speed in an observation time interval T, the traffic flow is termed homotachic. When the space and time headways are all equal in a homotachic regime, the flow is called *uniform* in T.

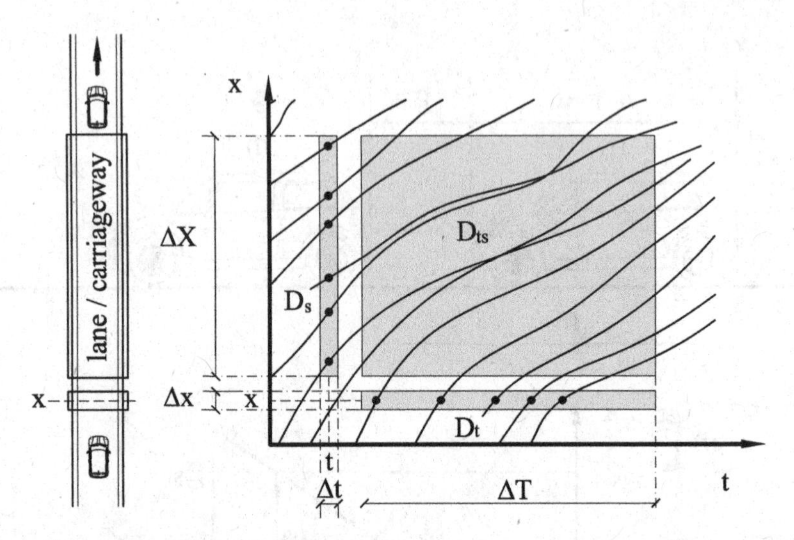

Fig. 1.2 Vehicle trajectories—space taken up in function of time (x = (t)), by a group of vehicles, and identification of potential domains (D_t, D_s, D_{ts}) for analysing the flows

1.2.1 *Traffic Variables Referred to the Time Domain*

Consider the time domain D_t. At the location x, for the time interval ΔT, the following traffic variables are defined [4]: the *flow rate*, the *time mean speed and the space mean speed.*

- The *flow rate* q(x) is the number of vehicles n(x) that pass x, in ΔT, expressed in vehicles per time unit:

$$q(x) = \frac{n(x)}{\Delta T} \tag{1.2}$$

Simply, the flow rate will be denoted with q(x).

Example 1 20 vehicles pass x in 40 s; the corresponding flow rate is q(x) = 20/40 = 0.5 veh/s = 1800 veh/h.

Consider now the relationship between the interval ΔT and the time headways h_{ti} of the vehicles that pass x in ΔT. Clearly, the higher the vehicle number n(x) in ΔT and the more accurately is: $\Delta T = \sum_{i=1}^{n(x)-1} h_{ti}$.

$\Delta T = \sum_{i=1}^{n(x)-1} h_{ti}$ is exact if the observation time period starts and ends immediately before the arrival of a vehicle.

Let $\overline{\tau}$ be the average time headway. $\overline{\tau}$ is:

$$\overline{\tau} = \frac{\sum_{i=1}^{n(x)-1} h_{ti}}{(n-1)} = \frac{\Delta T}{(n-1)} \tag{1.3}$$

With (1.2) from (1.3) the result is:

$$q(x) = \frac{n(x)}{[n(x) - 1] \cdot \overline{\tau}} \tag{1.4}$$

When n(x) increases, since [n(x) − 1] n(x), from (1.3) and (1.4) we have, respectively:

$$\overline{\tau} = \frac{\Delta T}{n} \tag{1.5}$$

$$q(x) = \frac{1}{\overline{\tau}} \tag{1.6}$$

For practical application, for traffic streams without platoons, Eq. (1.5) is useful approximation for the average time headway calculation.

Example 2 If the flow rate measured in a location x is 1200 vehicles/h, the average time headway estimated with (1.6) is:

$$\overline{\tau} = \frac{1}{q(x)} = \frac{3600\,\text{s/h}}{1200\,\text{veh/h}} = 3\,\text{s/veh}$$

which virtually coincides, by (1.3), with:

$$\overline{\tau} = \Delta T/[n(x) - 1] = 3600/1199 = 3\,\text{s/veh}$$

Example 3 As inferred from vehicle counts, in a given observation period the average vehicle headway is 4 s/veh; then, the corresponding flow rate is:

$$q(x) = \frac{1}{\overline{\tau}} = \frac{3600\,\text{s/h}}{4\,\text{s/veh}} = 900\,\text{veh/h}$$

Consider a mixed traffic stream (e.g., with $n_l(x)$ passenger cars, $n_m(x)$ motorcycles and $n_p(x)$ heavy vehicles). In order to assess the performance of a highway facility (in uninterrupted or interrupted flow) at different demand level, q(x) is expressed only in terms of passenger car unit (pcu): $q(x) = [n_l(x) \cdot c_l + n_m(x) \cdot c_m + n_p(x) \cdot c_p]/\Delta T$.

c_l, c_m and c_p are the *equivalence or homogenization coefficients.*

Equivalence coefficients are generally provided by level of service (LOS) evaluation procedures. They are function of highway geometry (e.g. vertical and horizontal alignment) and percentage of the different vehicle types, mainly.

Example 4 100 passenger cars, 100 motorcycles and 100 heavy vehicles pass a cross road section in an hour. Assume, on the basis of the specific situation, $c_l = 1$, $c_m = 0.5$ and $c_p = 2$, then the equivalent flow can be calculated in pcu/h: $q(x) = 100 \cdot 1.0 + 100 \cdot 0.5 + 100 \cdot 2 = 100 + 50 + 200 = 350$ pcu/h.

Time mean speed $\overline{v}_t(x)$ is the arithmetic average of the instantaneous speeds $v_i(x)$ of n(x) vehicles passing a given location x (spot speeds), during ΔT:

$$\overline{v}_t(x) = \frac{\sum_{i=1}^{n(x)} v_i(x)}{n(x)} \tag{1.7}$$

Simply, the time mean speed will be indicated with $\overline{v}_t(x)$ or with $\overline{v}_t$.

Example 5 During a time interval ΔT, 5 vehicles passing a location x with the instantaneous speeds $v_1 = 100$ km/h; $v_2 = 95$ km/h; $v_3 = 110$ km/h; $v_4 = 90$ km/h; $v_5 = 105$ km/h. The time mean speed results:

$$\overline{v}_t(x) = \frac{100 + 95 + 110 + 90 + 105}{5} = 100\,\text{km/h}$$

Space mean speed $\overline{v}_s(x)$ is the *harmonic mean* of the instantaneous vehicle speeds $v_i(x)$ measured at a given location x (spot speeds), during ΔT:

$$\overline{v}_s(x) = \frac{1}{\frac{1}{n} \cdot \sum_{i=1}^{n} \frac{1}{v_i(x)}} \tag{1.8}$$

Expression (1.8) has the meaning of travel mean speed. The following proof shows this meaning of Eq. (1.8).

Figure 1.3 shows a time domain $D_t \equiv (\Delta T, \Delta x)$. Let $\Delta t_1, \Delta t_2, \ldots \Delta t_n$ be the travel times of the vehicles 1, 2, ... n, travelling in ΔT, the segment Δx long.

The travel-mean time for the vehicles in Δt is:

$$\overline{\Delta t} = \sum_{i=1}^{n} \frac{\Delta t_i}{n} \tag{1.9}$$

So the travel-mean speed by Eq. (1.9) results:

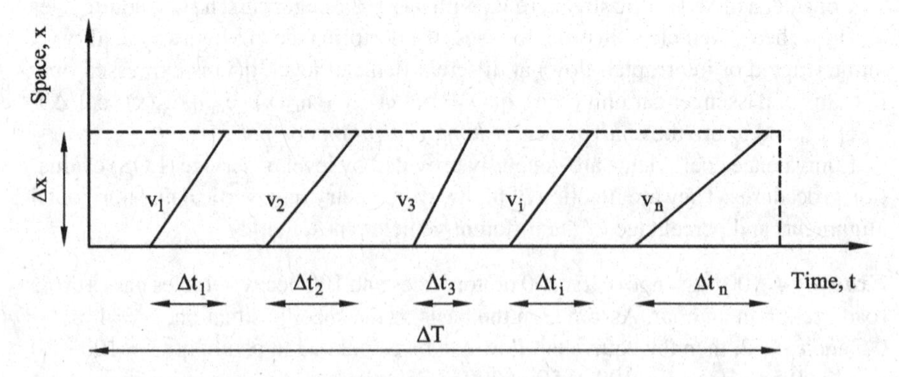

Fig. 1.3 Vehicle trajectories in an elementary time/space domain

$$\overline{v}_s(\Delta x) = \frac{\Delta x}{\overline{\Delta t}} = \frac{n \cdot \Delta x}{\sum_{i=1}^{n} \Delta t_i} \tag{1.10}$$

From Eq. (1.10), it results:

$$\overline{v}_s(\Delta x) = \frac{n}{\sum_{i=1}^{n} \frac{\Delta t_i}{\Delta x}} = \frac{1}{\frac{1}{n}\sum_{i=1}^{n} \frac{1}{v_i(\Delta x)}} \tag{1.11}$$

In Eq. (1.11) $v_i(\Delta x) = \frac{\Delta x}{\Delta t_i}$ is the travel speed of the vehicle i-th of Δt travelling Δx. Equation (1.10) is the harmonic mean of the n travel speeds in Δt.

If the segment Δx long is small, the instantaneous speeds of the n vehicle are virtually constant in D_t.

In other word, all the trajectories of the n vehicles in D_t can be considered virtually rectilinear (Fig. 1.3). Consequently a vehicle "i", i = 1, 2, … n traveling Δx in time Δt_i has a travel speed $v_i(\Delta x) = \Delta x/\Delta t_i$ equal to its instantaneous speed $v_i(x)$, in any location x of Δx, $v_i(\Delta x) = v_i(x)$.

Thus, Eq. (1.11) can be rewritten as:

$$\overline{v}_s(\Delta x) = \frac{1}{\frac{1}{n}\sum_{i=1}^{n} \frac{1}{v_i(x)}} \tag{1.12}$$

Equation (1.12) coincides with Eq. (1.11) so

$$\overline{v}_s(\Delta x) = \overline{v}_s(x) \tag{1.13}$$

Example 6 Consider the Fig. 1.4 where $\Delta T = 20$ s and $\Delta x = 50$ m.

The vehicles travel times are: $\Delta t_1 = 1.39$ s, $\Delta t_2 = 1.80$ s, $\Delta t_3 = 1.24$ s, $\Delta t_4 = 2.42$ s.

So, the travel speeds of the vehicles are:

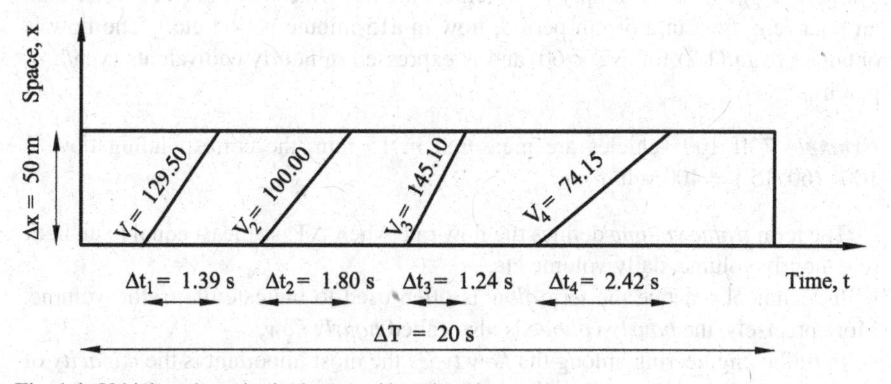

Fig. 1.4 Vehicle trajectories in the space/time domain

$\frac{\Delta x}{\Delta t_1} = 129.50\,\text{km/h}$; $\frac{\Delta x}{\Delta t_2} = 100.00\,\text{km/h}$; $\frac{\Delta x}{\Delta t_3} = 145.10\,\text{km/h}$; $\frac{\Delta x}{\Delta t_4} = 74.15\,\text{km/h}$.

The segment Δx long is so small to suggest that trajectories are rectilinear in every section of it and therefore the instantaneous vehicle speeds $v_i(t)$ remain constant in Δx. Thus, the travel speeds previously calculated coincide with the instantaneous ones in every location x of the segment.

Moreover, in the observation interval $\Delta T = 20$ s all the vehicles entering Δx, exit from it.

The space and time mean speeds can then be calculated on the same set of values $v_1(x)$, $v_2(x)$, $v_3(x)$, $v_4(x)$. With (1.12) we obtain:

$$\overline{v}_s(x) = \frac{4}{\left(\frac{1}{129.50} + \frac{1}{100.00} + \frac{1}{145.10} + \frac{1}{74.15}\right)} = 104.98\,\text{km/h}$$

On the other hand, the time mean speed is obtained by means of (1.7):

$$\overline{v}_t(x) = \frac{129.50 + 100.00 + 145.10 + 74.15}{4} = 112.19\,\text{km/h}$$

It goes without saying that the values of the two mean speeds do not coincide in that their expressions are different, even though they are calculated on the same data set. The constant flow in every section x of Δx is equal to:

$$q(x) = 4/20 = 0.2\,\text{veh/s} = 720\,\text{veh/h}.$$

1.2.2 Traffic Flow, Traffic Volume and Capacity

The term *traffic flow* or simply *flow* defines the flow rate when ΔT is shorter than an hour (e.g. flow in a 5-min period, flow in a15-minute period etc.). The flow is obtained from (1.2) for $\Delta T < 60'$ and is expressed in hourly equivalents (veh/h or pcu/h).

***Example** 7* If 100 vehicles are measured in 15 min, the corresponding flow is $100 \cdot (60'/15') = 400$ veh/h.

The term *traffic volume* defines the flow rate when ΔT is at least equal to an hour (e.g. hourly volume, daily volume etc.).

In technical practice the term *flow* is often used to indicate the traffic volume. More precisely, the *hourly volume* is also called *hourly flow*.

In traffic engineering among the *flow types* the most important is the *capacity* of a highway facility. For most applications the *capacity* is the maximum flow referred to a given period of time which has "sufficient probability of not being exceeded" in

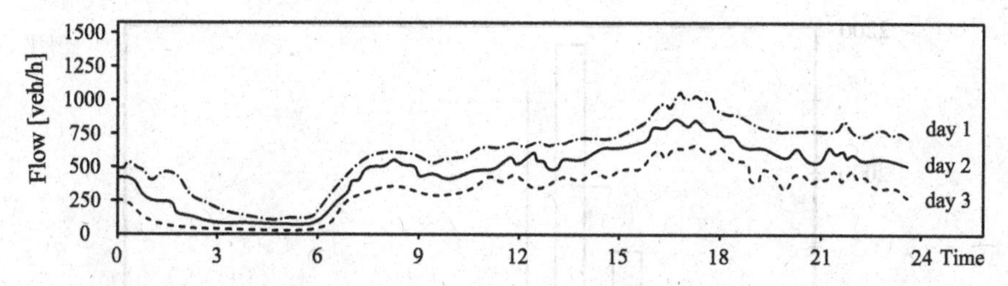

Fig. 1.5 Hourly volume variations on urban roads (example)

a cross road section or in a homogeneous segment of a lane or a carriageway. The *capacity* is estimated under prevailing roadway, traffic and control conditions [1].

Generally speaking, the capacity is also referred to a 15-min time interval, then extended to an hour and expressed in veh/h or pcu/h.

It is well-known that the traffic phenomena are random. Thus, the traffic volumes variations are *random processes* in space (along the road axis) and in time (hourly, daily, monthly and seasonal variations etc.). Then, considering, for instance, the hourly volume measured at a cross section of an urban road, the resulting hourly volume variations during the day are different day to day (see Fig. 1.5). Random and different from one another are also the volumes (and therefore their relevant flows) in sub-intervals of equal width in an hour.

Figure 1.5 shows the number of the passing vehicles at 5-min or 15-min intervals on a road cross section, however expressed in hourly equivalents (veh/h). Figure 1.6 shows traffic peaks and flow fluctuations more marked at 5 than 15 min. This is a quite general circumstance. The flows referring to traffic peaks, expressed in hourly equivalents, are called *peak flows*.

1.2.3 Peak Hour Factor

Experimentally the traffic congestion conditions are seen to derive from peak flows at 15-min intervals (see Fig. 1.6). Moreover, the 5-min or 10-min traffic peak effects on traffic state condition have lesser impact than 15-min traffic peaks.

Thus, it may happen that the peak volume is higher than the capacity "c" of a highway facility, even if the hourly volume, representing the traffic demand "d", is lower than "c" (see Sect. 2.1).

The flow may exceed the capacity only for a quarter of an hour, but the effects of the congestion in this peak period can even affect the whole hour and more.

Thus, when compared with the capacity of an highway facility (e.g. lane, roundabout entry, etc.), the peak flow at 15-min intervals in the analysed hourly volume DHV(x) is essential to facility design and measure of effectiveness (MOE) calculation. DHV(x) is termed *design hourly volume* in the observed cross section of abscissa x.

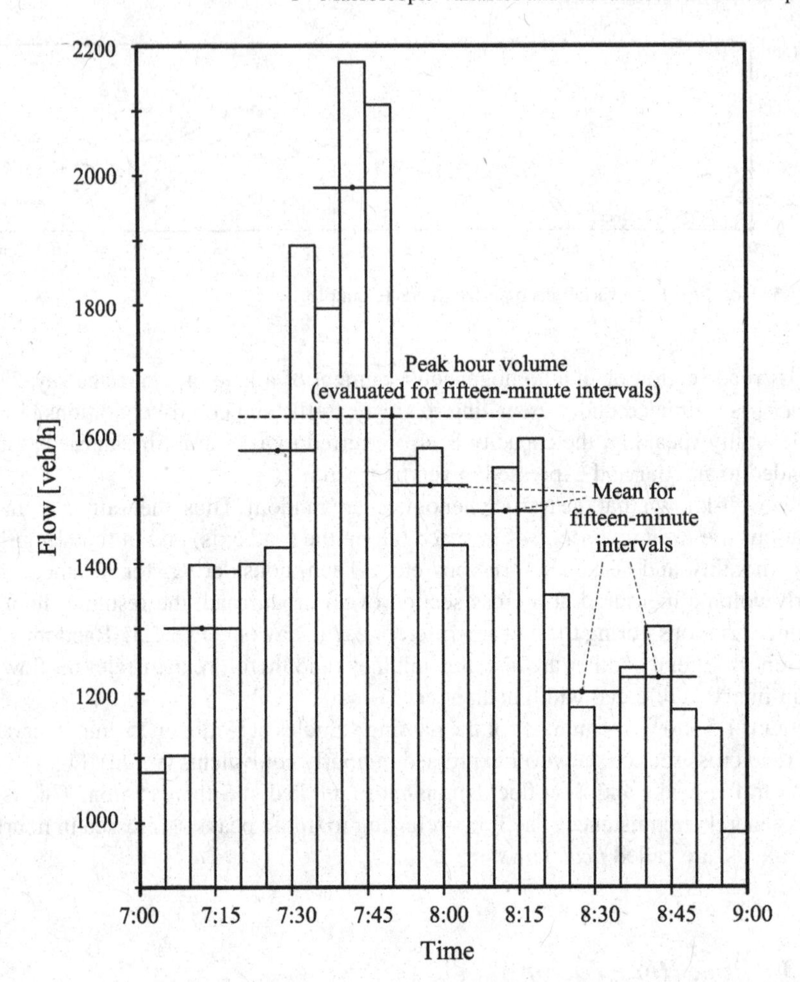

Fig. 1.6 Traffic volume variation over 5′ and 15′ (example) [1]

The peak flow referred to at 15-min intervals is called design flow $Q_p(x)$ or traffic intensity. So, in order to consider traffic flow fluctuations in the hour (see Fig. 1.5) and therefore potential consequent effects on a highway facility (e.g. oversaturation), the *peak hour factor* (PHF) is introduced:

$$\mathrm{PHF}(x) = \frac{\mathrm{DHV}(x)}{4 \cdot q_{15}(x)} \tag{1.14}$$

where $q_{15}(x)$ is the flow observed during the peak 15 min (veh/15′ or pcu/15′) in DHV(x). Therefore, $4 \cdot q_{15}(x)$ is the *design flow* $Q_p(x)$.

The design flow $Q_p(x)$ provides, in the light of what said above, the basis for highway facilities design and MOE calculation.

Should the peak hour factor PHF be known, the design flow $Q_p(x) = 4 \cdot q_{15}(x)$ can be obtained directly from the design hourly volume DHV(x) by (1.14):

$$Q_p(x) = \frac{DHV(x)}{PHF(x)} \tag{1.15}$$

PHF values can be clearly obtained from (1.15) by means of traffic surveys. In the absence of available traffic surveys, literature data on similar highway will be of help. The design hourly volume DHV(x), on the other hand, is selected according to the considerations in the section below.

Example 8 In the location x, in four consecutive quarters of an hour, there are measured: $q_1(x) = 25$ veh/15′; $q_2(x) = 75$ veh/15′; $q_3(x) = 50$ veh/15′; $q_4(x) = 100$ veh/15′. The peak flow is $q_4(x) = 100$ veh/h, while the design hourly volume DHV(x) (i.e. the sum of the four flows) is equal to 250 veh/h and the design flow $Q_p(x) = 4 \cdot q_4(x) = 4 \cdot 100 = 400$ veh/h.

Thus one has PHF(x) = 250/400 = 0.625.

1.2.4 Estimating of Design Hourly Volume DHV

Equation (1.15) shows that, should the peak hour factor PHF be known, an appropriate design hourly volume value DHV(x) needs to be set to calculate the design flow $Q_p(x)$. The DHV(x) value is usually obtained by processing hourly volume $q(x)_i$ recorded at a location x for a whole year.

Should the volume $q(x)_i$ of each of 8760 h in a year be measured *the cumulative frequency curve of hourly volume* can be drawn as in Fig. 1.7.

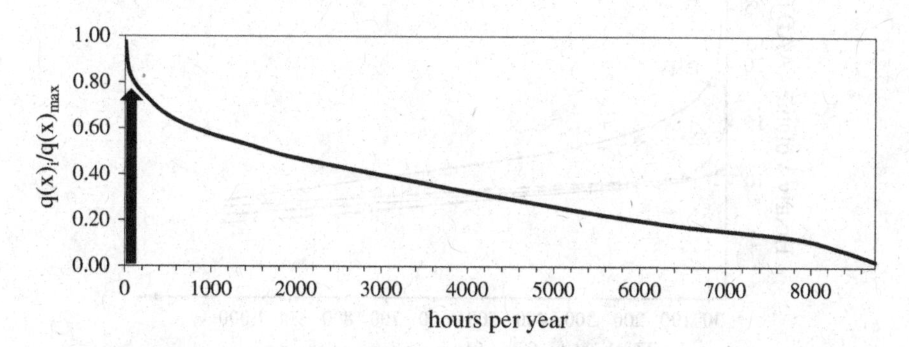

Fig. 1.7 Example of hourly volume frequency curve (the arrow qualitatively indicates the volumes assumed in the design and functional compliance phases: from the 30th to 100th peak hours of the year)

This diagram shows the hourly volume values on the ordinate axis and the number of the yearly hours on the abscissa axis (peak hour axis) in which the corresponding volume is reached or overreached.

The hourly volume corresponding to a number "n" of hours read on the abscissa is called volume of the n-th peak hour.

Figure 1.7 exemplifies a specific type of hourly volume cumulative frequency curve. In this case one has on the ordinate axis the ratio of the hourly volumes $q(x)_i$ to the highest hourly volume value $q(x)_{max}$ recorded in that year.

Figure 1.8 shows, for four different roads category, another type of hourly volume cumulative frequency curve. In this cases one has on the ordinate axis the ratio of the hourly volumes $q(x)_i$ measured to the corresponding annual average daily traffic AADT(x) value.

The Annual Average Daily Traffic (AADT) is a synthetic index widely used in Transportation Engineering. It is obtained by the ratio of the total number of vehicles n(x) passing a location x in a year to the number of the days in a year ($\Delta T = 365$ days). According to (1.2): $AADT(x) = n(x)/365$ [veh/day or pcu/day].

The curves typified in Figs. 1.7 and 1.8 can be assimilated to hyperbolic curves, in their turns similar to bilateral curves with vertex at around 30th peak hour of the year (see Fig. 1.8).

From the functional point of view, it should be advisable to choose the highest hourly volume recorded in the year as DHV(x) ($DHV(x) = q(x)_{max}$). This choice, however, leads to capacity over-dimensioning with respect to the prevailing traffic demand in the highway facility.

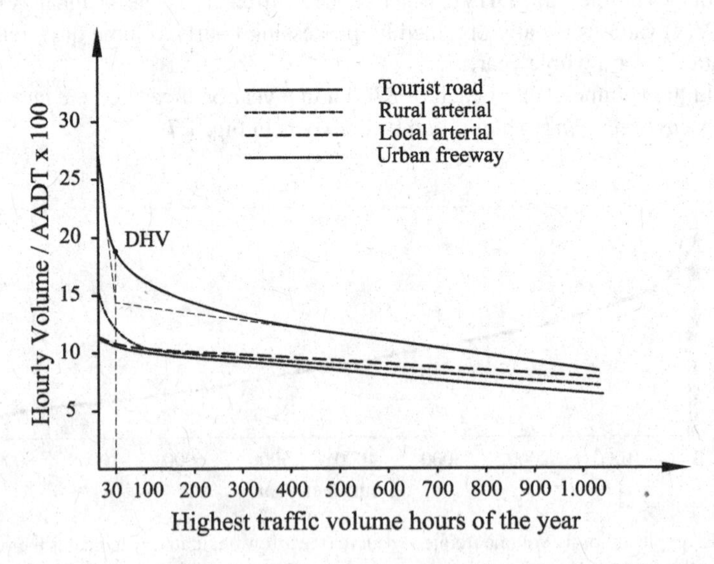

Fig. 1.8 Hourly volume/AADT for different road types and identification of the design hourly volume (DHV)

Table 1.1 Values of the coefficient α and peak hour factor (PHF)

Road types	α	PHF
Local arterial with no specific destination	0.12–0.15	0.80–0.90
Local arterial with elevated daily volumes	0.10–0.13	0.85–0.93
Local arterial with specific/tourist destinations	0.15–0.20	0.88–0.95
Urban roads with high traffic values	0.08–0.10	0.90–0.95
Freeways	–	0.85–0.90

A rational criterion for selecting the design hourly volume DHV(x) is inferred from observing the curves in Fig. 1.8. These curves show a marked variation of the slope between the 30th and 100th peak hours in the year.

If a highway facility is dimensioned by choosing a DHV*(x) value among those belonging to the steepest portion, it will appear markedly over-dimensioned for most hours of the year. In fact, for the remaining portion of the curve, the selected DHV(x) value will be significantly higher than traffic values observed in the largest part of the year. On the other hand, if the design hourly volume DHV**(x) selected for the facility designing ranges between the 30th and 100th peak hours of the year, probable congestion phenomena for a limited number of hours per year (that is, from 30 to 100 h) would be accepted; thus avoiding large over-dimensioning. In fact, with respect to the selected DHV(x) value, DHV**(x), the traffic values recorded in most of the year do not turn out to be significantly low as it would occur if a value were selected among those in the ascending portion of the cumulative frequencies curve.

For designing new highway facilities the design hourly volume DHV(x) is estimated with the models suitable for Transport Planning and Traffic Engineering.

The design hourly volume can be also obtained in function of the AADT through the relation:

$$\mathrm{VHP}(x) = \alpha \cdot \mathrm{AADT}(x) \tag{1.16}$$

Indicative values of the coefficient α and the peak hour factor (PHF), taken from [1] are shown in Table 1.1 for different road types.

1.2.5 *Traffic Variables Referred to the Space Domain*

Consider the space domain D_S. Along the segment ΔX long, at the time instant t (see Fig. 1.2), the *vehicle density* and the *space mean speed* are defined, as follows [4]:

- *vehicle density,* or simply, *density* $k_{\Delta X}(t)$, is the vehicle number m(t) on the segment ΔX long-expressed in vehicles per length unit- at t:

$$k_{\Delta X}(t) = \frac{m(t)}{\Delta X} \tag{1.17}$$

Simply the density will be indicated with k(t).

Example 9 From an aerial photo of a highway, taken at a generic time instant t, 15 vehicles are counted on a segment 500 m-long; the vehicle density results:

$$k(t) = \frac{15}{500} = 0.03\,\text{veh/m} = 30\,\text{veh/km}$$

Consider now the relationship between the segment ΔX long and the space headways hsi of the vehicles on this segment. Clearly, the higher the vehicle number m(t) in ΔX and the more accurately is $\Delta X = \sum_{i=1}^{m(t)-1} h_{si}$. $\Delta X = \sum_{i=1}^{m(t)-1} h_{si}$ is exact if the observation segment starts and ends immediately before of a vehicle. With (1.17) it results:

$$k(t) = \frac{m(t)}{\Delta X} = \frac{m(t)}{\sum_{i=1}^{m(t)-1m(t)-1} h_{si}} \tag{1.18}$$

When m(t) increases, since [m(t) – 1] m(t), from (1.18) we have:

$$k(t) \approx \frac{m(t)}{\sum_{i=1}^{m(t)} h_{si}} = \frac{1}{\frac{1}{m(t)}\sum_{i=1}^{m(t)} h_{si}} = \frac{1}{\sum_{i=1}^{m(t)} \frac{h_{si}}{m(t)}} = \frac{1}{\bar{h}_s} \tag{1.19}$$

Since $\bar{h}_s = \sum_{i=1}^{m(t)} \frac{h_{si}}{m(t)}$ is the mean space headway between vehicles along the segment ΔX long.

Example 10 If the vehicle density is k = 25 veh/km, the mean space headway between pairs of vehicles results as such:

$$\bar{h}_s = \frac{1}{k(t)} = \frac{1000\,\text{m/km}}{25\,\text{veh/km}} = 40\,\text{m/veh}$$

- *space mean speed* $\bar{v}_s(t)$ is the arithmetic average of the instantaneous speeds $v_i(t)$ of the vehicles that are on a segment ΔX long at t.

$$\bar{v}_s(t) = \frac{1}{m}\sum_{i=1}^{m} v_i(t) \tag{1.20}$$

where m is the number of vehicles that are on the segment segment ΔX long at time instant t.

Expression (1.20) has the meaning of travel-mean speed. The following proof shows this meaning of Eq. (1.19).

Figure 1.9 shows a space domain $D_s \equiv (\Delta t, \Delta X)$. Let $\Delta x_1, \Delta x_2, \ldots \Delta x_m$ be the distances travelled by the vehicles 1, 2, … m on the segment ΔX long in Δs.

The travel-mean distance for the vehicles in Δs is:

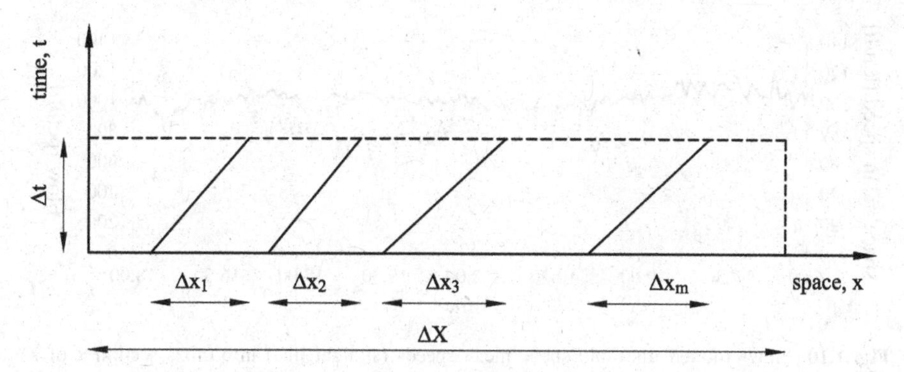

Fig. 1.9 Vehicle trajectories in an elementary space/time domain

$$\overline{\Delta x} = \sum_{i=1}^{m} \frac{\Delta x_i}{m} \tag{1.21}$$

So, the travel-mean speed results:

$$\overline{v}_s(\Delta t) = \frac{\overline{\Delta x}}{\Delta t} = \frac{1}{m}\sum_{i=1}^{m}\frac{\Delta x_i}{\Delta t} = \frac{1}{m}\sum_{i=1}^{m} v_i(\Delta t) \tag{1.22}$$

In Eq. (1.22) $v_i(\Delta t) = \frac{\Delta x_i}{\Delta t}$ is the travel speed of the i-th vehicle of Δs travelling Δx_i (i = 1, 2, … n).

Equation (1.22) is the arithmetic mean of the m travel speeds in D_s.

If the time interval Δt is small the instantaneous speed of the m vehicles are virtually constant in D_s. In other words all the trajectories of the m vehicles in D_s can be considered virtually rectilinear. Consequently, a vehicle "i" travelling Δx_i in the time Δt has a travel speed $v_i(\Delta t) = \frac{\Delta x_i}{\Delta t}$ equal to its instantaneous speed $v_i(t)$ in any time instant t of Δt.

So, Eq. (1.22) can be rewritten as:

$$\overline{v}_s(\Delta t) = \frac{1}{m}\sum_{i=1}^{m} v_i(t) \tag{1.23}$$

Equation (1.23) coincides with Eq. (1.20) therefore:

$$\overline{v}_s(\Delta t) = \overline{v}_s(t) \tag{1.24}$$

Traffic analyses systematically use the space mean speed. As later explained, under specific regular flow conditions, in case of uninterrupted-flow infrastructures, determinations on a few dozen metre-long segments, or the deduction starting from the instantaneous speeds in single segments, are sufficient, together with flow values, to characterize traffic conditions (see Fig. 1.10). For interrupted-flow infrastructures,

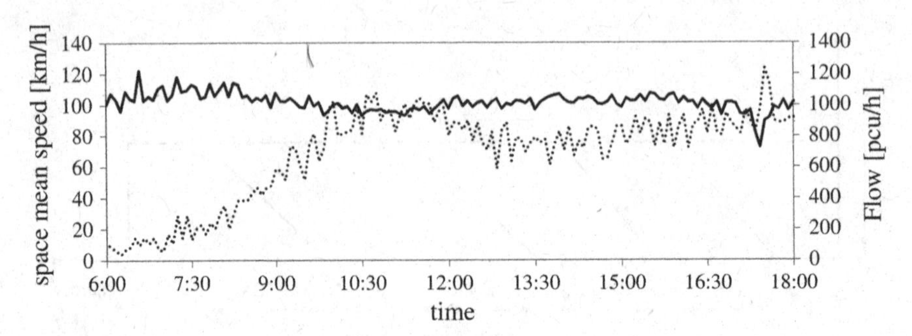

Fig. 1.10 Flows (dotted line) and space mean speeds (straight line) in a cross section x of a motorway lane (A22, Italian motorway)

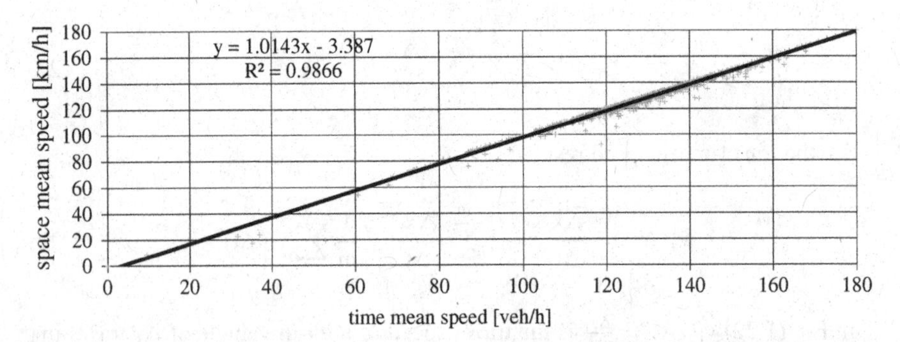

Fig. 1.11 Example of a relation between time mean speed and space mean speeds in a motorway lane (A22, Italian motorway)

on the other hand, segments should be long enough to encompass the elements which contribute more than others - as a result of the stops there implemented - to increase travelling times and decrease space mean speeds.

Therefore, time mean speeds are not generally significant for analysing interrupted flow infrastructures.

Figure 1.11 displays a typical relation between $\overline{v}_s(x)$ (Eq. (1.8)) and $\overline{v}_t(x)$ (Eq. (1.7)) derived from processing experimental data. As explained in Sect. 1.2.6, this is clearly a general result (see Eq. (1.33)).

1.2.6 Fundamental Flow Relationship

The fundamental flow relationship links the macroscopic flow variables, i.e. flow, density and space mean speed, to each other with a state equation. The fundamental flow relationship can be achieved in several ways. A simple and easily understandable inference of the physical meaning of this relationship is given below. Consider the

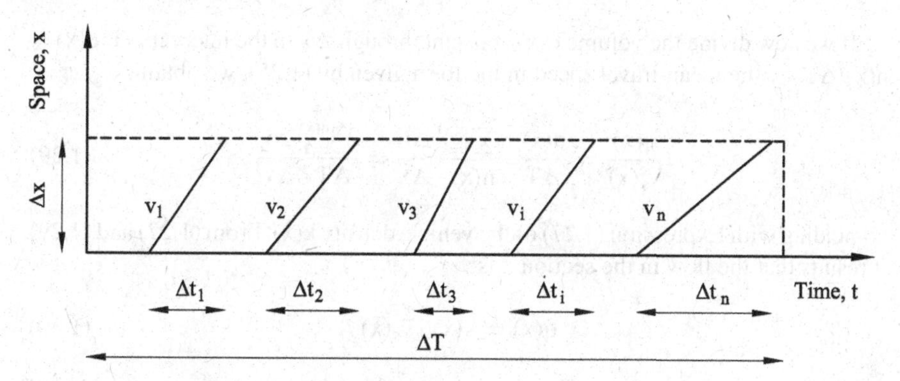

Fig. 1.12 Vehicle trajectories in an elementary time/space domain

generic domain D_t in Fig. 1.3 and replaced in Fig. 1.12 All the n vehicles entering the segment Δx long travel it in ΔT. In other word, all the n vehicles entering this segment exit from it in ΔT. This is realistic because Δx is small. So, the flow is $q(x) = n(x)/\Delta T$. For the n vehicles in ΔT the space mean speed from Eq. (1.13) and Eq. (1.10) results:

$$\overline{v}_s(x) = \frac{n \cdot \Delta x}{\sum_{i=1}^{n} \Delta t} \tag{1.25}$$

Again with reference to a time domain D_t, the mean vehicle density $k_{\Delta T}(t)$ on segment Δx long can be determined and calculated for each instant t of ΔT.

In order to calculate this density, the number of times which every vehicle i – i = 1, 2, … n—is present in the domain D_t, is measured by the interval Δt_i taken by the vehicle "i" during ΔT travelling the segment Δx long.

Thus, in the observation interval ΔT the measurement of the vehicle presence in the segment Δx long is given, on average, by [5]:

$$\Theta = \frac{\sum_{i=1}^{n(x)} \Delta t_i}{\Delta T} \tag{1.26}$$

Θ is called *occupation.*

For the density on the segment Δx long (straddling the road cross section with abscissa x) thus we have:

$$\overline{k} = \frac{\Theta}{\Delta x} = \frac{\sum_{i=1}^{n(x)} \Delta t_i}{\Delta T \cdot \Delta x} \tag{1.27}$$

Since the mean density so determined concerns a domain D_t, it is indicated as:

$$k(x) = \overline{k} = \frac{\sum_{i=1}^{n(x)} \Delta t_i}{\Delta T \cdot \Delta x} \tag{1.28}$$

If we now divide the volume q(x) constant through Δx in the interval ΔT, $q(x) = n(x)/\Delta T$, by the mean-travel speed in the form given by (1.25), we obtain:

$$\frac{q(x)}{\overline{v}_s(x)} = \frac{n(x)}{\Delta T} \cdot \frac{\sum_{i=1}^{n(x)} \Delta t_i}{n(x) \cdot \Delta x} = \frac{\sum_{i=1}^{n(x)} \Delta t_i}{\Delta T \cdot \Delta x} \tag{1.29}$$

coinciding with Expression (1.27) of the vehicle density k(x). From (1.27) and (1.29) it results that the flow in the section x is:

$$q(x) = k(x) \cdot \overline{v}_s(x) \tag{1.30}$$

where $\overline{v}_s(x)$ is (cfr. Eqs. (1.12) and (1.13)):

$$\overline{v}_s(x) = \frac{1}{\frac{1}{n}\sum_{i=1}^{n} \frac{1}{v_i(x)}} \tag{1.31}$$

in which $v_i(x)$, $i = 1, 2, \ldots n$, is the instantaneous speed of the vehicles in ΔT, in any location x of the segment ΔX long.

Formula (1.30) is the fundamental flow relationship referred to the road sections x, or to the elementary segment Δx straddling x observed during the elementary time interval Δt.

For the sake of simplicity, (1.30) is written as:

$$q = k \cdot v \tag{1.32}$$

where v denotes the space mean speed.

Example 11 With the data from Example 6 (Fig. 1.4) with Eq. (1.30) we obtain:

$$k(x) = q(x)/\overline{v}_s(x) = 720/104.98 = 6.8\ \text{veh/km}$$

The density previously calculated coincides with the value obtained by means of Eq. (1.28):

$$k(x) = \frac{\sum_{i=1}^{n(x)} \Delta t_i}{\Delta T \cdot \Delta x} = \frac{(1.39 + 1.80 + 1.24 + 2.42)}{20 \cdot 50} = 0.0068\ \text{veh/m} = 6.8\ \text{veh/km}$$

1.2.7 *Traffic Flow Stationarity*

Equation (1.30) is an in-mean relation and represents the basis for *deterministic flow models* which will be described in Chap. 2.

In view of what has been achieved, Eq. (1.30) has a local meaning where the space mean speed $\overline{v}_s(x)$ is the harmonic mean of instantaneous speeds in an elementary segment of the flow observation section.

Equation (1.30) cannot then be used for analysing a significantly long road segment, except under special statistical regularity conditions in traffic.

In order to quantitatively specify these conditions, it is necessary to remind of the links between the time mean speed $\overline{v}_t(x)$ and the space mean speed $\overline{v}_s(x)$. This is due to the fact that $\overline{v}_t(x)$ and $\overline{v}_s(x)$ are an arithmetic mean and a harmonic mean respectively. For the relations between these types of mean when done on the same speed sample v_i of numerousness "n", it results [6, 7]:

$$\overline{v}_t(x) = \overline{v}_s(x) + \frac{s_s^2(x)}{\overline{v}_s(x)} \tag{1.33}$$

$$\overline{v}_s(x) = \overline{v}_t(x) - \frac{s_t^2(x)}{\overline{v}_t(x)} \tag{1.34}$$

where $s_s^2(x)$ and $s_t^2(x)$ are respectively the variances of speeds $v_i(x)$ calculated with regard to the means $\overline{v}_t(x)$ and $\overline{v}_s(x)$:

$$s_s^2(x) = \frac{1}{n}\sum_{i=1}^{n(x)}[v_i(x) - \overline{v}_s(x)]^2 \tag{1.35}$$

$$s_t^2(x) = \frac{1}{n}\sum_{i=1}^{n(x)}[v_i(x) - \overline{v}_t(x)]^2 \tag{1.36}$$

Between variances there is the following relation:

$$s_t^2(x) = \frac{s_s^2(x)}{\overline{v}_s(x)} \cdot \overline{v}_t(x) \tag{1.37}$$

which is immediately obtained from (1.33) and (1.34).

Example 12 With the data from Example 6 (Fig. 1.4) with (1.36) we achieve $\overline{v}_t(x) = (129.50 + 100.00 + 145.10 + 74.15)/4 = 112.19$ km/h.

$$s_t^2(x) = \frac{1}{4}[(129.50 - 112.19)^2 + (100.00 - 112.19)^2 + (145.10 - 112.19)^2 + (74.15 - 112.19)^2] = 744.58\ (\text{km/h})^2$$

With (1.34) we get the following estimation:

$$\overline{v}_s(x) = 112.19 - 744.58/112.19 = 105.55\ \text{km/h}$$

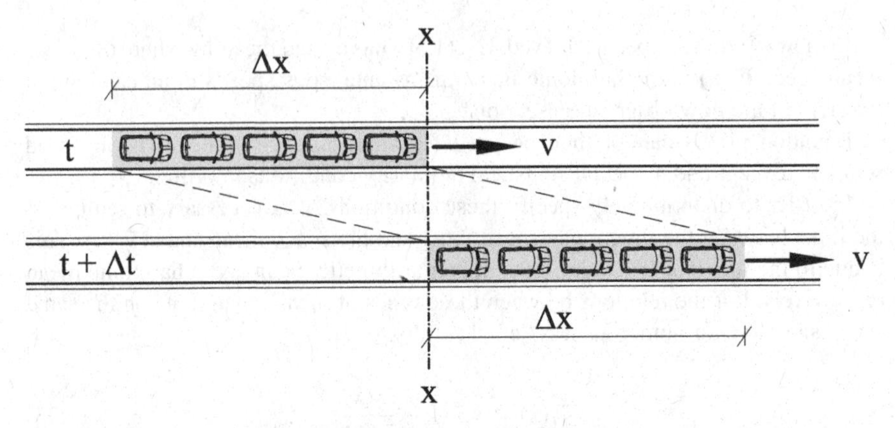

Fig. 1.13 Hydrodynamic analogy as explanation of uniform flow stationarity. The coloured area contains $m = k \cdot \Delta x$ vehicles. In the time interval $\Delta t = \Delta x/v$, the vehicles all pass through on the cross section x. The flow is $q = m/\Delta t = k \cdot \Delta x/\Delta t = k \cdot v$.

with respect to the value, calculated with (1.8) directly from the data, $\overline{v}_s(x) = 104.98$ km/h, as shown in Example 6.

Equations (1.33) and (1.34) provide the more accurate the estimations, the higher the numerousness n of the data sample.

The fundamental flow relationship ((1.30) or (1.32)) appears to be particularly useful in order to define *stationarity* for technical purposes [8, 9]. It indicates that the space mean speed—in other words, the harmonic mean of instantaneous speeds—is known when the mean and variance of the instantaneous speeds are determined by means of (1.7) and (1.36) respectively.

An interrupted flow is said to be stationary when, in the presence of constant flow, the statistical distribution of instantaneous speeds has the same mean and variance in all the road sections.[1]

From (1.34) follows the space invariance of the space mean speed and from the fundamental flow relationship (1.32) follows the vehicle density invariance, in other words the distribution homogeneity of vehicles all along the infrastructure.

As regards the significance attributed to density, it involves uniformity in traffic conditions in terms of mutual conditioning between vehicles and users' psychological comfort levels.

In short, a traffic stream is stationary along a section L and during a period T if in each section x of L the traffic variables q, k, v do not vary during T and if these values are the same for each section x of L. Figure 1.13 shows the particular case of *uniform flow stationarity*. The uniform flow is defined above in Sect. 1.2. This example, even

[1]In terms of probabilistic characterisation of the flow phenomenon, the flow stationarity derives from, on the one hand, the stationarity-in-mean of the counting process due to the flow q being constant, and on the other, the stationary in-mean and in-variance (weak stationarity) of the instantaneous speeds in road sections. The two processes need being analysed to verify the statistical properties stated above, in a given time interval T along the infrastructure axis (in space).

if specifically referred to special conditions of flow regularity, can further clarify the stationarity concept.

The strict definition of flow stationarity requires traffic processes to be characterised as random processes [10]. This topic will be dealt with in Chap. 5. It needs to be underlined that the flow stationarity is a purely theoretical condition and it can be observed only approximately in real-life situations; it is however applied very often in building traffic models, which are highly useful in practical applications.

References

1. Highway Capacity Manual: HCM (2016) Transportation Research Board, Washington, DC
2. Papageorgiou M (1991) Concise encyclopedia of traffic & transportation. Pergamon Press (1991)
3. Rogers M, Enright B (2017) Highway engineering. Wiley, 3th edn
4. Buisson C, Lesort JP (2010) Comprendre le trafic routier: methodes et calculs. CERTU
5. Salter RJ (1989) Highway traffic analysis and design, 2nd edn. Palgrave Macmillan, UK
6. Wardrop JG (1952) Some theoretical aspects of road traffic research. In: Proceedings of the institution of civil engineers, vols 1 and 2
7. Gerlough DL, Huber MJ (1975) Traffic flow theory; a monograph. Special report 165. TRB
8. Ni D (2016) Traffic flow theory. Elsevier
9. Hoogendoorn SP, Traffic flow characteristics. Delft University of Technology. https://ocw.tudelft.nl/.
10. Mauro R (2015) Traffic and random processes. Springer

Chapter 2
Macroscopic Traffic Flow Models

Abstract This chapter offers an intuitive definition of transport demand, capacity and flow. The main macroscopic flow models, levels of service and traffic hysteresis phenomena are also dealt with.

Traffic flow models are mathematical relationships between traffic variables. They are used to describe the vehicle flow and predict its development over time.

The adopted approach will lead to formulate deterministic and/or stochastic models. The *deterministic models* can be of three types: macroscopic, microscopic and mesoscopic.

Macroscopic models are based on relationships between the following flow variables: flow q(x), space mean speed $\overline{v}_s(x)$ and density k(x), defined in Chap. 1.

Microscopic models study the interactions between single vehicles which follow one another in a traffic stream. Microscopic models are introduced in Chap. 4.

Mesoscopic models schematize traffic streams as a sequence of vehicle packets properly charaterised kinematically. Mesoscopic models are touched upon in Chap. 4.

Stochastic models are based on the description of traffic processes over time, that is, as random processes. They are dealt with synthetically in Chap. 5.

This chapter focuses on the main macroscopic flow models and their main properties.

2.1 Transportation Demand, Capacity and Flow

In Chap. 1 traffic flow and capacity are introduced as proper mean values (see Eq. 1.2 and Sect. 1.2.2). Among the other variables used in traffic studies, the traffic demand is certainly of interest.

For road networks the traffic demand d(O → D) between an origin O and a destination D is the vehicles number moving from O to D in a reference time unit. The reference time unit is usually set as equal to one hour.

M. Guerrieri and R. Mauro, *A Concise Introduction to Traffic Engineering*,
Springer Tracts in Civil Engineering,
https://doi.org/10.1007/978-3-030-60723-4_2

Thus the transportation demand, traffic flow and capacity on road network links are expressed in veh/h or pcu/h. It is generally achieved with the methods of the Transport System theory [1, 2].

There are clear relationships between the demand $d(O \rightarrow D)$, capacity $c(x)$ and flow $q(x)$.

The following example illustrates some aspects of these relationships.

Example 1: Consider the case in Fig. 2.1a.

Two roads xz and yz, with two lanes each, converge in a main three-lane segment zh. The capaciy for each lane is 2000 veh/h on average. For segments xz and yz the

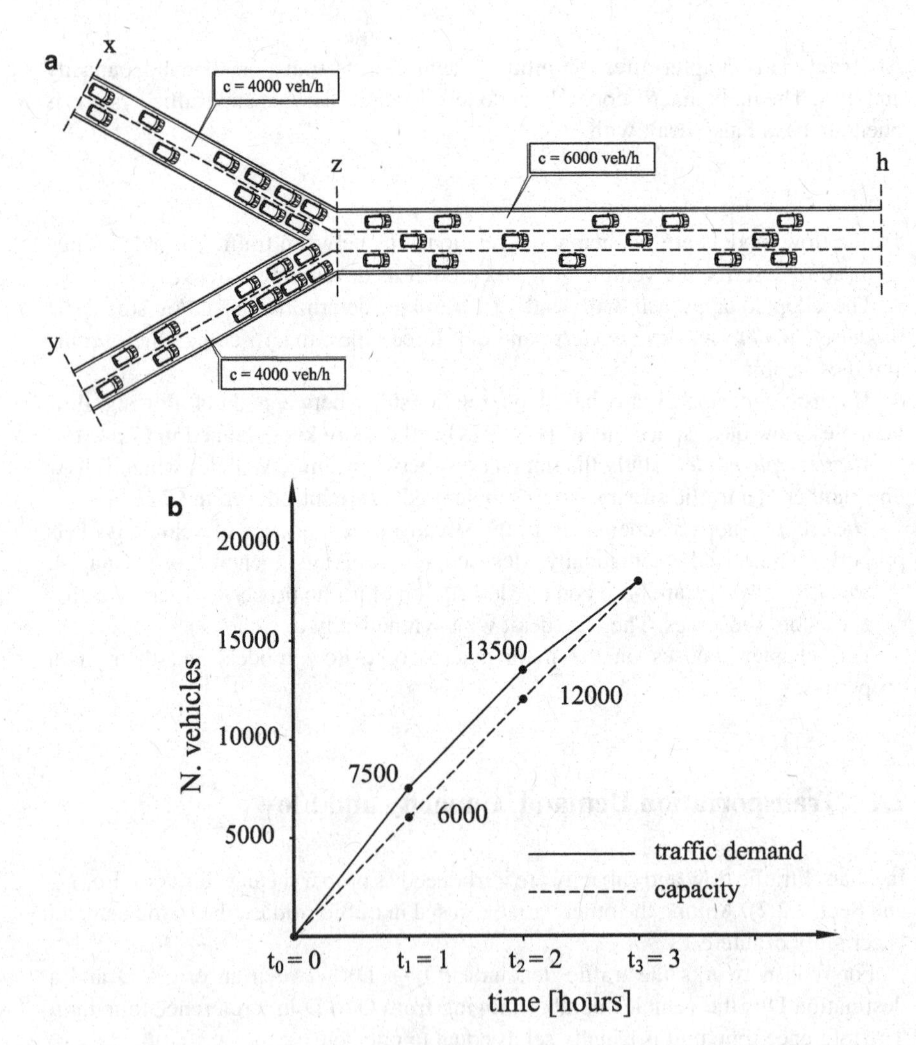

Fig. 2.1 **a** Confluence of two side streets (xz and yz) into a main road (zh). **b** Traffic demand, capacity and number of waiting vehicles in the time interval t_0–t_3

capacity is then 2·2000 = 4000 veh/h. For the main segment the traffic capacity is equal to 3·2000 = 6000 veh/h.

For the system the traffic demand is initially lower than the capacity in segments xz, yz and zh.

Thus, the flow of the latter segments runs smoothly without any queue formation.

Afterwards (Fig. 2.1b), starting from t_0 (conventionally $t_0 = 0$) to $t_1 = 1$ h, the vehicles moving from x to h (demand $d(x \rightarrow h)$) and those moving from y to h (demand $d(y \rightarrow h)$) increase over time. The increase occurs at constant average rates $d(x \rightarrow h) = 3800$ veh/h and $d(y \rightarrow h) = 3700$ veh/h.

All the values calculated in the following are to be meant as mean values.

Therefore, in the time interval $t_0 - t_1$ the total demand flow towards the segment z, entering the main segment, is 7500 veh/h (3800 + 3700 veh/h).

As that flow is greater than the capacity c(z) in z (c(z) = 6000 veh/h), the flow running from the segment z is equal to the capacity q(z) = c(z) = 6000 veh/h. On the two segments xz and yz, upstream of z, the forming queues grow steadily and continuously in the time interval $t_0 - t_1$.

In short, at t_1, after an hour from t_0, the maximum total number of queuing vehicles $\overline{N}(t)$ (sum of the queuing vehicles on xz and yz) is:

$$\overline{N}(t) = q(z) - c(z) = 7500 - 6000 = 1500 \text{ veh.}$$

Starting from $t_1 = 1$ h the total traffic demand through z decreases to $d(x \rightarrow h) + d(y \rightarrow h) = 6000$ veh/h. It remains steady for another hour up to $t_2 = 2$ h. The flow running through z is equal to the capacity in z, q(z) = c(z) = 6000 veh/h.

Thus, in $t_1 - t_2$ the total number of vehicles queuing on the segments xz and yz remains unchanged and is equal to $\overline{N}(t) = 1500$ veh. After t_2, $\overline{N}(t)$ decreases progressively since the total demand towards z decreases from the average value of 6000 veh/h to that of 4500 veh/h. This demand is lower than the capacity c(z) in the segment z. Therefore, the flow q(z) running from z is equal to the traffic demand q(z) = 4500 veh/h.

The queues on the two segments xz and yz disappear completely ($\overline{N}(t_3) = 0$ veh) at time $t_3 = \overline{N}(t_2)/[q(z) - c(z)] = 1500/(6000 - 4500) = 1$ h, after three hours from t_0.

This example was developed by applying the deterministic queuing theory. The results turned out to be the more realistic the more limited the variations in the traffic demand, capacity and flows over the analysed time interval, with respect to their averages. The *deterministic queuing theory* will be dealt with in Chap. 8.

2.2 Relationship Between Macroscopic Traffic Flow Variables

Experimentally, in uninterrupted flow conditions there are relationships among the following variables: flow q, density k and space mean speed v, defined in Chap. 1. In fact, when a road section is travelled by few vehicles fairly distant from each other, the vehicle density is very low as well as the flow in a road section. In these so-called *free flow conditions*, every user maintains the *desired speed*. Therefore, the flow speed is high on average, unless there are unfavourable conditions in the geometrical alignment, environment etc.

Starting from the free flow, a progressive increase of the traffic demand in the section entails an increase in the running flow. It follows a greater and greater reduction in vehicle headways and consequently an increase in vehicle density. Drivers, being mutually conditioned, are increasingly less free of reaching the desired speed. This causes a progressive decrease in the average flow speed. In fact, when overtaking becomes more difficult and riskier, the slowest vehicles influence the speeds of the other vehicles in the traffic flow. This leads to platoon formation behind the slowest vehicles and then to progressive reduction in the average speed along the road segment. Figure 2.2 exemplifies an experimental evidence of the traffic behaviour described so far. Figure 2.2 displays the performance of observation data points selected in a section of a freeway lane.

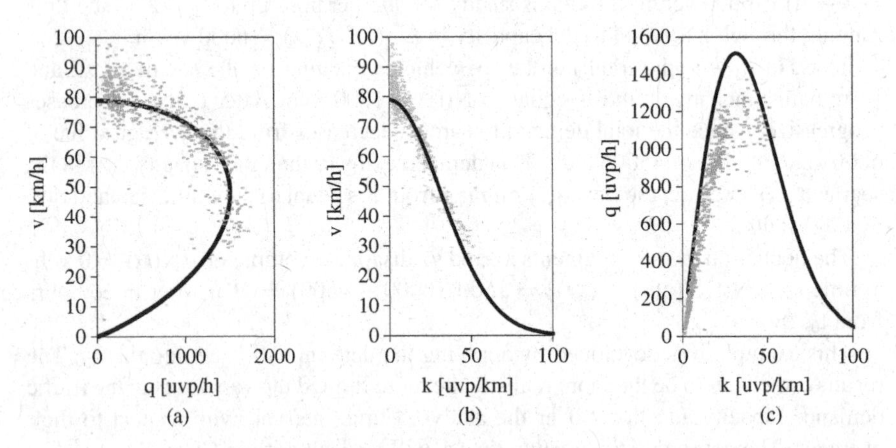

Fig. 2.2 Example of q, k and v measurements in a freeway lane cross section and relationships v = v(q), v = v(k), q = q(k), interpolated from the observation data points by least squares method

2.3 Properties of the Mathematic Traffic Flow Models and Operational Flow Conditions

In this and in the following Sections, for the sake of simplicity, it is assumed q(x) = q, k(x) = k, $\overline{v}_s(x) = v$.

The fundamental flow equation (Eq. 1.32) worked out in Chap. 1 is:

$$q = k \cdot v \tag{2.1}$$

Equation (2.1) represents a hyperbolic paraboloid in a three-dimensional space.

The physically significant area of the surface (2.1) is that contained in the first octant (Fig. 2.3), in that q, k and v values are always positive or null.

If Eq. (2.1) is associated to any of the three experimental relations v = v(q), v = v(k), q = q(k), it yields a system of two linear equations in three unknowns. This system, implicitly, identifies a curve C on the space (q, k, v) which lies on the surface of Fig. 2.3. Moreover, if one of the three relations in (2.1) is given, the other two relations of the model are obtained univocally (see Sect. 2.9). For instance, if the relation v = v(k) is known, from (2.1) the relations v = v(q) and q = q(k) are univocally obtained.

Geometrically speaking, the curves v = v(q), v = v(k) and q = q(k) are the three orthogonal projections onto the coordinate planes q-v, k-v and k-q of the curve C. The set {v = v(q), v = v(k), q = q(k)} is a macroscopic (or in-mean-field) model of the traffic flow.

Figure 2.4 exemplifies the three projections of the curve C concerning the *Greenshields traffic stream model.*

The representation in Fig. 2.4 allows the diagrams v = v(q), v = v(k) and q = q(k), and the necessary relations in mean among v, q and k to be concurrently displayed.

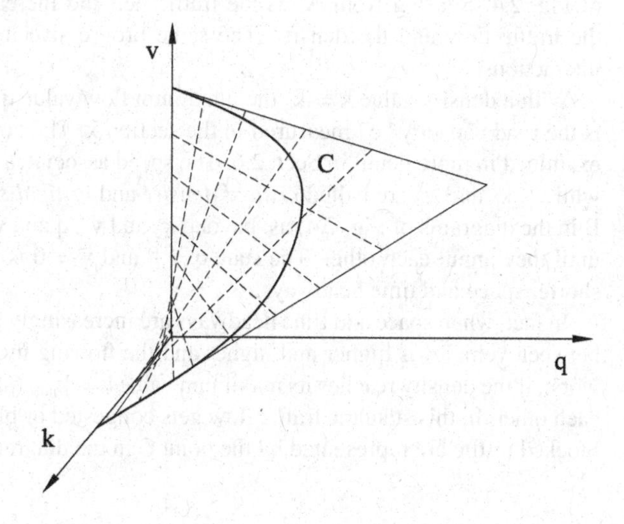

Fig. 2.3 Example of a curve (q; v; k)

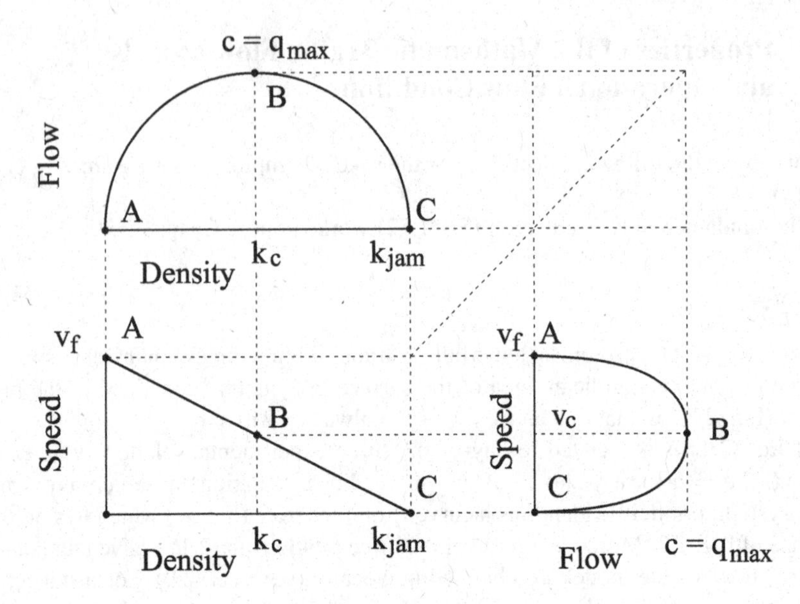

Fig. 2.4 Projection of in-mean relation among q, v, k onto coordinate planes

In order to be close to the real traffic phenomenon, these relations must satisfy some fundamental conditions.

Notably, in agreement with experimental observations (Sect. 2.2), when the flow is free, the time and space headways among vehicles are very high and the conditioning between vehicles is extremely low.

In that case (see Eqs. 1.6 and 1.12) the traffic flow q and the density k are near zero, and the space mean speed tends to the so-called *free flow speed* v_f.

The completely free flow conditions are presented by the point "A" in the diagrams of Fig. 2.4. Starting from A, as the traffic demand increases progressively, so do the traffic flow and the density. The same progressive increase occurs in vehicle interactions.

With a density value $k = k_c$ the maximum flow value $q = q_{max}$ is recorded. q_{max} is the road capacity "c" measured in the section x. The concept of capacity will be examined in more detail in Sect. 2.6. The speed associated to k_c and q_{max} is denoted with v_c. k_c and v_c are called *critical density* and *critical speed,* respectively (point B in the diagrams of Fig. 2.4) as, beyond k_c and v_c, q and v decrease monotonically until they annul each other. The state $q = 0$ and $v = 0$ is reached with shorter and shorter space and time headways.

In fact, when space and time headways are increasingly reduced, the conditioning between vehicles is higher and higher and the flowing more and more difficult. At worst, if the density reaches its maximum value $k = k_{jam}$ (*jam density*), q and v annul each other. In this state the traffic flow gets congested or blocked. The conditions of blocked traffic are represented by the point C in the diagrams of Fig. 2.4. By way of

Table 2.1 Values of the flow parameters estimated for an Italian motorway [3]

Lane/carriageway	v_f (km/h)	k_c (pcu/km)	c (pcu/h)	v_c(km/h)
Driving lane	107	24	1.540	65
Overtaking lane	129	25	1.950	78
Carriageway	116	48	3.370	70

exemplification, Table 2.1 shows the values of the flow parameters estimated for an Italian motorway.

2.3.1 Single-Regime Models

The trends in the graphs illustrated in Fig. 2.4 are typical of the Greenshields traffic stream model [4]. In fact, in the Greenshields model v = v(k) is linear, v = v(q) is a horizontal-axis parabola and q = q(k) is a vertical-axis parabola (Table 2.2).

The *Greenshields model* belongs to single-regime models, in that v = v(q), v = v(k), and q = q(k) are each a single continuous function.

Table 2.2 reports the relations v = v(q), v = v(k), and q = q(k) of two other single-regime models.

Generally based on experimental observations, *Drake's model* is valid for roads characterised by medium–low density values.

Greenberg's model is valid for roads characterised by medium–high density values. The Greenshields model is still today one of the most used in technical applications for its semplicity.

Table 2.2 Single-regime models

Model name (Year)	v = v(k)	q = q(v)	q = q(k)	Model parameters
Greenshields (1935)	$v = v_f \cdot \left(1 - \frac{k}{k_{jam}}\right)$	$q = k_{jam} \cdot \left(v - \frac{v^2}{v_f}\right)$	$q = v_f \cdot \left(k - \frac{k^2}{k_{jam}}\right)$	v_f; k_{jam}
Drake, Schofer, May (1967)	$v = v_f \cdot e^{-\frac{1}{2}\left(\frac{k}{k_c}\right)^2}$	$q = k_c \cdot v \cdot \left(-2 \cdot \ln \frac{v}{v_f}\right)^2$	$q = v_f \cdot k \cdot e^{-\left(\frac{k}{k_c}\right)^2}$	v_f; k_c
Greenberg (1959)	$v = -v_0 \cdot \ln\left(\frac{k}{k_{jam}}\right)$	$q = k_{jam} \cdot v \cdot e^{-\left(\frac{v}{v_0}\right)}$	$q = -v_0 \cdot k \cdot \ln\left(\frac{k}{k_{jam}}\right)$	v_0; k_{jam} (with v_0 speed for $k \longrightarrow 0$)

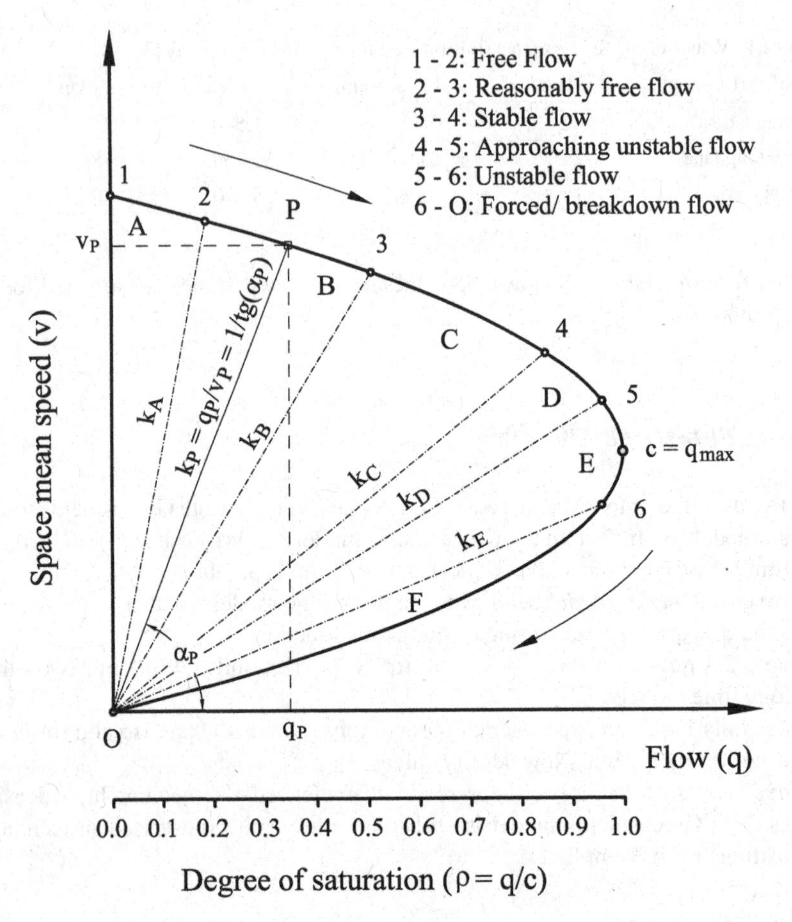

Fig. 2.5 The Greenshield relation v = v(q) and subdivision of the curve in function of flow conditions

For the different single-regime models refer to [5] in which they are described and compared in detail.[1]

In [6] further meaningful comparisons are introduced between single-regime models.

The v = v(q) diagram of a single-regime model provides very valuable information for applications. Figure 2.5 displays a typical Greenshields' traffic flow diagram. The conditions of free (1–3), stable (3–4) and near-stable (4–5) flows are described from the upper part of the parabola v = v(q). The unstable regime (5–6) sets around the capacity value $c = q_{max}$. Such a regime (5–6) is comprised of partly the upper segment

[1]The models of Underwood (1961), Northwestern (1967), Drew (1968), Newell (1961), Pipes-Munjal (1967), Greenshields modified (1995), Del Castillo and Benitez (1995), Van Aerde (1995), MacNicholas (2008) are all single-regime models [7].

and partly the first part of the lower segment of v = v(q). The constrained flow is described from the lower segment (6-O) of v = v(q).

In qualitative terms, a *traffic flow is stable* if a perturbation in the flow regime does not cause a significant reduction in speed or traffic flow.

Perturbations have manifold causes: for instance, rapid increases in traffic demand, slowing downs or frequent lane changes of vehicles in the stream.

The free flow has the highest stability degree.

A *traffic flow is unstable* even if small perturbations cause systematic rapid speed drops and, in some cases, also the complete traffic stop.

The *forced flow* represents the highest instability degree.

In near-instable situations, the stable-flow conditions alternate unexpectedly to the unstable-flow ones with shorter or longer periods of time.

2.3.2 *Levels of Service*

Figure 2.5 shows the operational traffic conditions, also denoted with letters from A to F. According to the terminology in the Highway Capacity Manual (HCM) [8], the states from A to F are called "levels of service" (LOS).

The *level of service* is a measure of traffic flow quality which corresponds to an exact traffic flow value.

LOS are calculated for given road, traffic, user and environment characteristics.

For uninterrupted-flow roads, the traffic flow conditions which characterize each level of service, are (see Figs. 2.5 and 2.6):

- *A—free flow* with no mutual vehicle conditioning: very low vehicle density, very high driving comfort;

Fig. 2.6 Examples of levels of service [8]

- *B—stable flow*: low density which imposes relatively rare restrictions on freedom of manoeuvre; however, drivers always enjoy considerable freedom of choice in speed as well as in driving lanes; phase B is still in reasonably free flow conditions with high driving comfort;
- *C—still a stable flow*: increase in density values; greater conditioning between vehicles. Fairly frequent lane-changing and/or overtaking in order to maintain the desired speed; the average speed attained is, however, still satisfactory; still acceptable driving comfort;
- *D—approaching unstable flow*: high density and reduced freedom of manoeuvre; still mostly stable operational conditions; acceptable average speeds even if strongly influenced by changes in traffic flow conditions; poor driving comfort;
- *E—unstable flow*: nearly the same traffic flow as the capacity value; very high density; virtually zero driving freedom; any perturbation in the traffic flow brings about rapid speed drops to complete standstill; extremely poor driving comfort;
- *F—forced or breakdown flow*: density at its highest values; vehicles move up in the queue under stop and go conditions; by far the poorest standards in driving conditions.

The separation limits between the different levels of service vary with road type. Based on LOS, every road is characterised by specific intervals of the variable describing the traffic flow.

In case of motorways and two or multilane highways, the levels of service (LOS) depend on one only parameter, the density k (Fig. 2.6).

The different LOS are then defined by consecutive k intervals.

For Eq. (2.1) it results $k = q/v$. Therefore, a segment OP between the origin O of the q-v axes and a point P (q_P, v_P) on the curve $v = v(q)$ is inclined by $1/k_P = tg(\alpha_P)$ with respect to the q axis (see Fig. 2.5).

Figure 2.5 shows the segments from O to be inclined with respect to the q axis on the basis of the k values defining the levels of service.

The intersection of these segments splits the upper part of $v = v(q)$ curve into arches corresponding to different levels of service. Thus, LOS are also defined by intervals of q and v values.

If the traffic flow is known, speed and the corresponding level of service are univocally determined. For instance, in Fig. 2.5 the point P on the curve $v = v(q)$ corresponds to q_p. The speed v_P and the level of service B correspond to point P.

The q intervals defining LOS can also be standardised in relation to the c capacity. Therefore, the LOS are also obtained according to the *saturation degrees* $\rho = q/c$, $\rho \in [0, 1]$.

LOS identification according to the k density based on HCM 2016 is shown in Table 2.3 and Fig. 2.7. Figure 2.7 illustrates five $v = v(q)$ relations with a different performance from the Greenshields model (see Fig. 2.5).

These relations refer to five free-flow speeds for basic freeway segments under *ideal conditions*. As shown in Fig. 2.7, unlike Fig. 2.5, the lower limit of LOS E is near the capacity.

Table 2.3 Levels of service for rural and suburban multilane highways

Level of service (LOS)	Density k (pcu/km/lane)
A	0–7
B	>7–11
C	>11–16
D	>16–22
E	>22–28
F	Demand exceeds capacity or k > 28

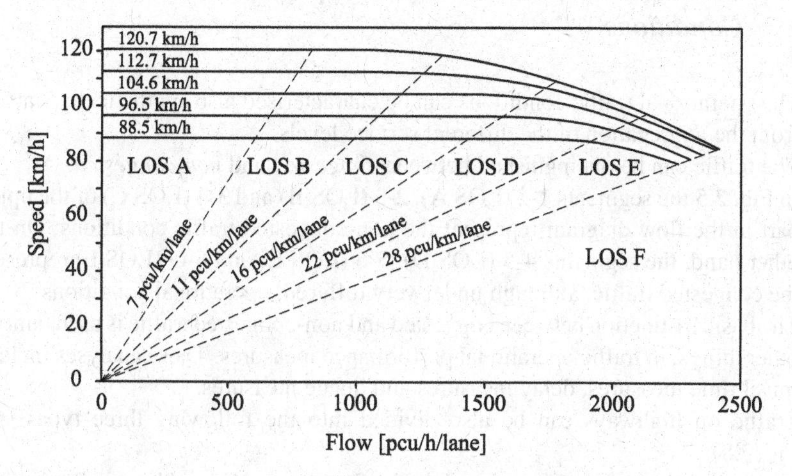

Fig. 2.7 LOS identification according to the density k based on HCM 2016 [8]

For *two-lane highways* LOS is not calculated on the basis of k but rather on average driving speed and the time wasted while waiting for overtaking opportunities.
The capacity and the level of service obtained at a given flow for each road depend on different factors. The main geometric elements taken into consideration are the *planimetric and altimetric alignment* and the *road* section (carriageway width, lane width, etc.). Environmental factors encompass weather conditions and the light (day or night). Traffic factors concern the flow conditions in terms of real vehicles number and type. As far as users are concerned, they are subdivided into habitual and non-habitual drivers.
In order to generalize the procedure for the LOS calculation, the HCM method defines the flow variable values for each LOS under ideal conditions.
The ideal conditions for each road type described in the HCM (two-lane highways, freeways, multilane highways, etc.) refer to the road geometry, environmental characteristics, user type and traffic. With regard to the latter condition, traffic variables are always expressed only in terms of passenger car units (i.e. q in pcu/h; k in pcu/mi/ln).

The HCM procedure also takes into account the varied real situations by means of proper corrective coefficients. This method is common to the capacity manuals of other countries.
Thus, when applied to real cases traffic flows turn into equivalent only-car streams. The transformation is carried out with the coefficients introduced in Sect. 1.2.1. As regards the HCM-based applications, see the latest edition [8].

2.3.3 *Other Characterizations of the Operational Traffic Conditions*

The operational traffic conditions can be characterised also synthetically starting from the aggregation of the different service levels.
The traffic can be distinguished between *congested* and *non-congested*.
In Fig. 2.5 the segments 1-2 (LOS A), 2-3 (LOS B) and 3-4 (LOS C) of the upper part in the flow diagram represent the non-congested traffic conditions. On the other hand, the segments 4-5 (LOS D), 5-6 (LOS E) and 6-O (LOS F) represent the congested traffic, although under very different operational conditions.
The basic distinction between congested and non-congested traffic is used, among other things, in traffic operational performance measures. These analyses include travel time measures, delay measures and queue measures.
Traffic on highways can be also divided into the following three types (see Fig. 2.8):
• *undersaturated flow*: it occurs when the flow is not affected by any bottleneck, upstream and downstream of the flow;
• *queue discharge flow*: it occurs when the flow overcomes a bottleneck and accelerates to reach the desired free flow speed. In the absence of other bottlenecks, the queue discharge flow is fairly stable until the queue is totally over.

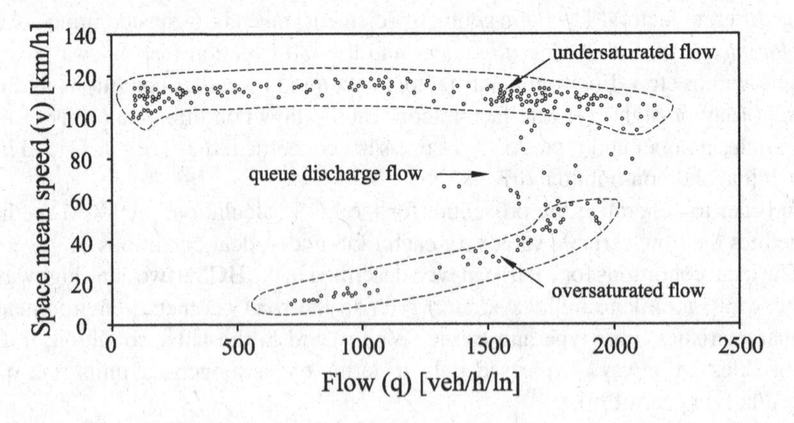

Fig. 2.8 The three types of flow in highways and motorways

• *oversaturated flow*: it occurs when flow conditions are affected by queues at bottlenecks. These conditions do not depend on the (prevailing) infrastructure characteristics but rather on flow problems (e.g. flow instability, see Chap. 4). Oversaturated flows are to be considered congested.
The scattering of the measures (q_i; v_i) in Fig. 2.8 shows a recent tendency of the traffic behaviour in motorways and highways. In effect, v in the stable regime tends to vary slightly when q increases. This tendency is well represented by the shape of the curve $v = v(q)$ in Fig. 2.7 provided by the HCM 2016 [8].

2.4 Experimental Definition of a Traffic Flow Model

Different devices and different operational modalities [9] are available in order to obtain accurate measures of q, k and v in a road cross section of abscissa x (see Sects. 1.2.1 and 1.2.5).
The flow parameters which can be observed directly and simply are flows and instantaneous speeds.
Direct density measures are, on the other hand, more complex because they require more sophisticated instruments and data processing [9].
Therefore, in order to obtain the diagrams $v = v(q)$, $v = v(k)$ and $q = q(k)$ the procedure is such that k is indirectly obtained by flow and speed measures.
During single time intervals Δt_i (e.g. 5 or 15 min), the transit vehicles number and instantaneous vehicles speed are measured.
For each interval Δt_i Eq. (1.2) allows to calculate the q flow espressed in veh/h or homogenized in passenger cars pcu/h (Sect. 1.2.1), and Eq. (1.13) allows to calculate the space mean speed v.
The measure pair (q_i; v_i) is associated to every Δt_i. The measures are reiterated for N intervals Δt_i, all of the same width.
N must be properly large to significantly sample the different traffic situations in the observated road section.
Starting from the measure pairs (q_i; v_i), $q = k \cdot v$ (Eq. 1.20) allows to rapidly calculate the density k_i referring to the interval Δt_i. Through k_i further N pairs (k_i; v_i) and N pairs (k_i; q_i) related to N observation intervals Δt_i are formed.
The pairs (k_i; v_i) reported on a Cartesian axes system, density-speed, lead to a *scatter plot*. These pairs are preferred in specifying the relationship between macroscopic variables, in the form $v = v(k)$ (see Table 2.2). $v = v(k)$ is calibrated with the *least squares method*. Thus one of the three relations of the experimental flow model is obtained.
The other two relations can be deduced by associating $v = v(k)$ to the relation $q = k \cdot v$ (Eq. 2.1) (see Sects. 2.9, 2.10 and 2.11).
The curve $q = q(k)$ is called *fundamental diagram of traffic*.
While the curve $v = v(q)$ is called *traffic flow diagram*.
In general, the identification of a deterministic traffic flow model from a certain set of experimental data is not univocal (see Sect. 2.5).

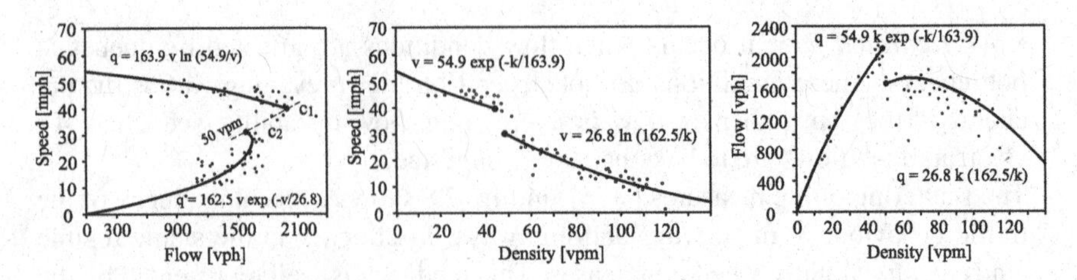

Fig. 2.9 Edie's two-regime relations

As usually with mathematical models, the selection of the most suitable model for experimental data handling is suggested by practical convenience.
The goodness of fitting between the adopted model and experimental data can be evaluated with a *conformity statistical test*.

2.5 Deterministic Multi-regime Traffic Flow Models

The previous section has introduced some macroscopic deterministic single-regime traffic flow models (see Table 2.2).
Multi-regime flow models are, on the other hand, those whose relation between q, k and v is represented by more functions. Moreover, the curves $v = v(k)$, $v = v(q)$ and $q = q(k)$ in most cases appear to be discontinuous against capacity and critical speed values.
Generally, *two-regime models* are taken into consideration. Figure 2.9 illustrates *Edie's two-regime model* [10] which interpolates some experimental data.
In Fig. 2.9 the two capacity values c_1 and c_2 are clearly achieved starting from different flow conditions:
• c_1 is reached from stable flow conditions (upper segment of the relation between v and q);
• $c_2, < c_1$ concerns the vertical upflow from forced flow conditions (lower segment of the relation between v and q).
Two-regime models were developed in an attempt to provide the closest description of the flow phenomenon to real-life flow conditions.
A broad array of two-regime models is provided in [5].

2.6 Physical Interpretation of the Capacity

The capacity of a road lane or carriageway is defined for technical purposes as the maximum flow able to run along a lane or carriageway section.
Another valid definition for applications, and equivalent to the previous one, is that the *capacity is the flow value that is quite likely not to be overcome in a given road*.

The two definitions are qualitative and conventional.
Experience suggests that the capacity ranges within very high value limits in function of varied factors. Such factors mainly depend on vehicle flow characteristics which are basically determined by drivers' behaviours. In their turn, drivers' behaviours depend on several causes, such as environmental conditions, travelling reasons, road geometric characteristics etc.
In road designing, however, a reference capacity value needs to be assumed for each different road type. This value is set equal to the average of the maximum traffic volumes effectively measurable in prevailing road type situations.
The average is worked out quite approximately by the maximum flow value of the traffic flow diagram v = v(q) (or also the diagram q = q(k)) which interpolates data pairs (q_i, v_i) (or (q_i, k_i)).
The pairs (q_i, v_i) or (q_i, k_i) are detected on a road section in different traffic situations (see Sects. 2.3 and 2.5).
Consequently, the traffic flow diagram v = v(q) and the fundamental traffic flow diagram q = q(k) are higly important in Highway Engineering, although they are all in-mean models that do not describe the dynamic nature of the flow phenomenon (see Chap. 3).
On the other hand, *Dynamic flow models* are necessary for setting up a proper traffic control strategy (see Chap. 6).

2.7 Hysteresis in Traffic Flows

Generally speaking, Edie's fundamental diagram q = q(k) (Fig. 2.9) can well be approximated by a bilateral curve, whose shape is a *reversed lambda* [11–13] (Fig. 2.10a).
The simplified representation of q = q(k) in Fig. 2.10 (*reversed* λ) is useful for analysing the capacity discontinuity or drop and hysteresis phenomenon in traffic flows.

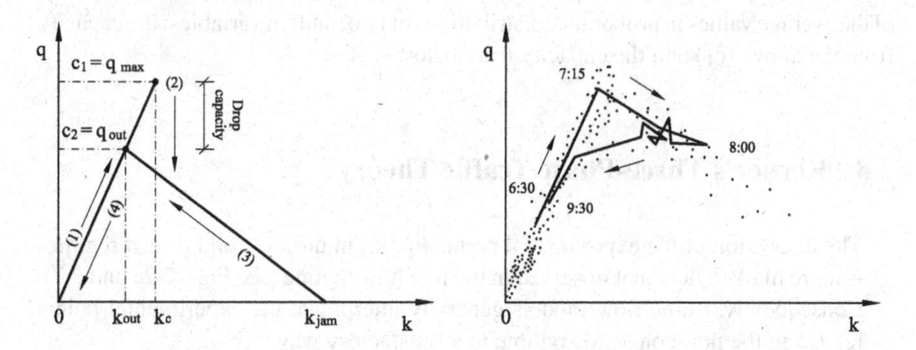

Fig. 2.10 Drop capacity (**a**) and hysteresis cycle (**b**) in the traffic flow diagram

In Fig. 2.10a in stable flow conditions, the traffic flow q increases when the density k rises (towards 1). q is equal to the traffic demand affecting the road segment under analysis.
When the density k tends to the critical density value k_c, traffic flow approaches instability until it becomes instable near k_c, reaching the capacity c_1.

This value c1 is said to be the capacity "from above".
It is observed empirically that for $k > k_c$ there is a steep fall in traffic flow which becomes unstable (towards 2). In this state the traffic demand is lower than the flow.
When traffic conditions evolve towards increasing k values, the flow becomes forced and the running flow gradually decreases.
Once achieved the forced regime, any decrease in traffic demand leads to decreases in k. If at this phase $k < k_c$, the system does not trace back its trajectory. The flow evolves towards 3. Only below a certain threshold k_{out}, further decreases in demand (towards 4) result in pairs (q_i, k_i) placed on the stable flow (towards 1).
In short, the transition from unstable to stable flow (due to decreases in demand) occurs at a density k_{out} and a flow q_{out} (queue discharge capacity).
k_{out} and q_{out} are respectively lower than k_c and c_1 which discriminate the transition from stable to unstable flow.
$q_{out} = c_2$ identifies the secondary capacity value. c_2 is achieved by the system when the traffic stream goes upwards from the forced to stable flow.
For this reason c_2 is said to reach the capacity "from below".
Thus the sequence identifies a hysteresis cycle in the fundamental flow diagram (from 1 to 4, through 2 and 3), which can be better represented by Fig. 2.10b referring to time progression (6:30–9:30 a.m.).
Figure 2.10b shows the temporal sequence of the pairs (q_i; k_i) before, during and after a traffic demand peak in the morning.
Thus the two capacity values c_1 and c_2, previously indicated in the two-regime models (Sect. 2.5), are clearly highlighted.
However, a reversed λ model does not allow the random nature of the two capacity values c_1 and c_2 to be seen.
In fact, c_1 and c_2 derive from two regression curves and thus they are an estimation of the average values in probability distributions of two random variables: the capacity from the above (c_1) and the capacity from below (c_2).

2.8 Kerner's Three-Phase Traffic Theory

The dispersion of the experimental points (q_i; k_i) in unstable and forced regimes is more marked than that observed in the free flow regime (see Figs. 2.2c and 2.7). Consequently, traffic flow models generally interpolate the experimental points related to the non-congested regime in a satisfactory way.

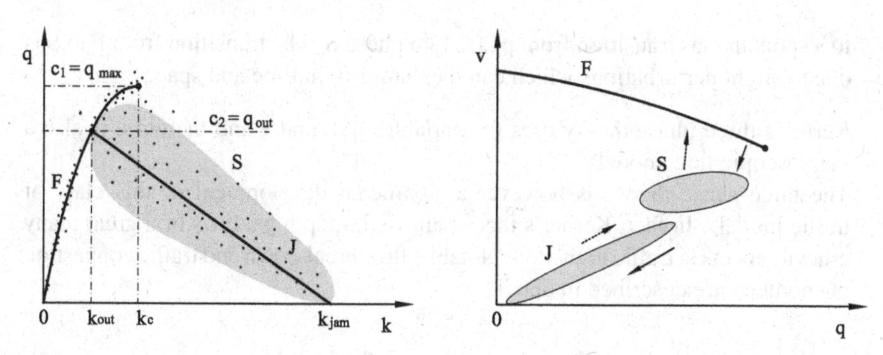

Fig. 2.11 Diagrams q = q(k) and v = v(k) in Kerner's three-phase traffic theory and transition among the three phases

Experimental studies carried out by Kerner [14] on the German highway A5 (with three lanes for each travelling direction) have allowed the development of the so-called "*three-phase theory*".

According to the three-phase theory, the functional relation between flow and density $q = q(k)$ is not valid for the entire density domain ($0 \leq k \leq k_{jam}$). The relation $q = q(k)$ is significant only for the non-congested regime ($0 \leq k \leq k_c$). Especially, in agreement with this theory, the fundamental diagram (see Fig. 2.11) shows:

- a curve in the free flow regime;
- a region or domain representing the dispersion of the values q-k in the congested regime.

Kerner's terminology is not that traditionally used in traffic engineering, where the free flow concerns the LOS A (see Fig. 2.5).

Therefore, according to Kerner three different phases can be identified in a traffic flow diagram (Fig. 2.11):

- *free flow* "F": the average speed on the overtaking lane is higher than that in the neighbouring lane. In its turn, the latter speed is higher than the speed in the right lane (when present);
- *synchronized flow* "S": density increases and average vehicle speeds on the carriageway lanes tend to assume comparable values, unlike what occurs in the free flow;
- *wide-moving jam* "J": it concerns areas with high density and negligible speed. The areas are delimited by an initial section (head) and a final section (tail). Head and tail move over time. Near these sections, flow speed values appear to be discontinuous, that is, the speed decreases in vehicles approaching the queue section of the congested area. J is characterized by a speed increase in vehicles leaving that area. Once overcome the head, speeds tend to the desired free flow speed.

The immediate transition from phase F to phase J does not occur spontaneously. It is triggered by marked flow perturbations (e.g. accidents). On the other hand, the generation of *traffic flow breakdown phenomena* (see Chap. 4) are correlated to a spontaneous transition from phase F to phase S. The transition from F to S is due to slight perturbations which can then amplify in time and space.

Kerner's three-phase theory uses the variables q, k and v and therefore is also a macroscopic flow model.
The three-phase theory is however a significant development of this class of traffic models. In fact, Kerner's theory allows in-depth analysis of a great many crucial aspects in traffic highways. Notably, flow breakdown and traffic congestion phenomena are described in detail.

2.9 Case Study: Service Levels According to Greenshields' Model

Experimental analyses have indicated the following value intervals for the flow conditions below:
- free flow: $0 < k_1 \leq 0.05\ k_{jam}$
- reasonably free flow: $0.05\ k_{jam} < k_2 \leq 0.14\ k_{jam}$
- stable flow: $0.14\ k_{jam} < k_3 \leq 0.30\ k_{jam}$
- approaching unstable flow: $0.30\ k_{jam} < k_4 \leq 0.38\ k_{jam}$
- unstable flow: $0.38\ k_{jam} < k_5 \leq 0.60\ k_{jam}$
- forced or breakdown flow: $0.60\ k_{jam} < k_6 \leq 1.00\ k_{jam}$.

These intervals are different than those in the HCM manual (Fig. 2.7).
Moreover, the following Greenshields' relation between the space mean speed (v) and the density (k) is known: $v = 75 - 0.60\ k$.
The aim is to find the flow condition corresponding to the flow $q = 1500$ veh/h and the saturation degree.
In order to solve the problem, the first step is to determine the jam density k_{jam} which is calculated from the previous expression for $v = 0$:
$0 = 75 - 0.60\ k_{jam}$ $k_{jam} = 75/0.60 = 125.00$ veh/km/lane.
The highest density interval values for every flow condition are:
$k_1 = 0.05\ k_{jam} = 0.05 \cdot 125.00 = 6.25$ veh/km/lane
$k_2 = 0.14\ k_{jam} = 0.14 \cdot 125.00 = 17.50$ veh/km/lane
$k_3 = 0.30\ k_{jam} = 0.30 \cdot 125.00 = 37.50$ veh/km/lane
$k_4 = 0.38\ k_{jam} = 0.38 \cdot 125.00 = 47.50$ veh/km/lane
$k_5 = 0.60\ k_{jam} = 0.60 \cdot 125.00 = 75.00$ veh/km/lane
$k_6 = 1.00\ k_{jam} = 1.00 \cdot 125.00 = 125.00$ veh/km/lane
The space mean speeds (v) and flows (q) corresponding to the above mentioned density values are:
$v_1 = 75 - 0.60\ k_1 = 75 - 0.60 \cdot 6.25 = 71.25$ km/h
$v_2 = 75 - 0.60\ k_2 = 75 - 0.60 \cdot 17.50 = 64.50$ km/h
$v_3 = 75 - 0.60\ k_3 = 75 - 0.60 \cdot 37.50 = 52.50$ km/h
$v_4 = 75 - 0.60\ k_4 = 75 - 0.60 \cdot 47.50 = 46.50$ km/h

$v_5 = 75 - 0.60\ k_5 = 75 - 0.60 \cdot 75.00 = 30.00$ km/h
$v_6 = 75 - 0.60\ k_6 = 75 - 0.60 \cdot 125.00 = 0.00$ km/h
$q_1 = v_1 \cdot k_1 = 71.25 \cdot 6.25 = 445.31$ veh/h
$q_2 = v_2 \cdot k_2 = 64.50 \cdot 17.50 = 1128.75$ veh/h
$q_3 = v_3 \cdot k_3 = 52.50 \cdot 37.50 = 1968.75$ veh/h
$q_4 = v_4 \cdot k_4 = 46.50 \cdot 47.50 = 2208.75$ veh/h
$q_5 = v_5 \cdot k_5 = 30.00 \cdot 75.00 = 2250.00$ veh/h
$q_6 = v_6 \cdot k_6 = 0.00 \cdot 125.00 = 0.00$ veh/h
Free flow speed v_f is determined for $k = 0$. Thus, it results:
$v_f = 75 - 0.60\ k = 75 - 0.60 \cdot 0 = 75$ km/h.
Considering the Greenshields relation $q = q(k)$ (see Table 2.2) and that $k_{jam} = 2 \cdot k_c$ (see Fig. 2.4), thus the flow ($c = q_{max}$) is:
$c = q_{max} = (v_f \cdot k_{jam})/4 = (75 \cdot 125.00)/4 = 2344.00$ veh/h.
The saturation degrees are equal to:
$\rho_1 = q_1/c = 445.31/2344.00 = 0.19$
$\rho_2 = q_2/c = 1128.75/2344.00 = 0.48$
$\rho_3 = q_3/c = 1968.75/2344.00 = 0.84$
$\rho_4 = q_4/c = 2208.75/2344.00 = 0.94$
$\rho_5 = q_5/c = 2250.00/2344.00 = 0.96$
$\rho_6 = q_6/c = 0.00/2344.00 = 0.00$
For $q = 1500$ veh/h the saturation degree is $\rho = 1500/2344 = 0.64$.
Finally, by solving the expression $q = k_{jam} \cdot (v - v^2/v_f)$ with regard to speed, the result is $v = 60.32$ km/h. Moreover, the flow is stable as shown in Fig. 2.12.

2.10 Case Study: Calibration of Greenshields' Flow Model

In monitoring a motorway tunnel, one of the activities is to determine the flow law in the travelling lane starting from the data obtained through inductive loops in the time 7:00 to 8:30 a.m.

It is well worth noting that the case under study only serves an illustrative purpose. As a matter of fact, in real-life cases, in order to calibrate the flow model, traffic data sampling should refer to a really huge observation hour number and in any case such to make it possible to identify different flow states (from free to congested).

The inductive loop records the time instant for every vehicle passage and its instantaneous speed (v_i). Data were collected in intervals of $\Delta T = 5$ min, as exemplified in Table 2.4 for the interval 7:00:00–7:05:00.

For this interval $\Delta T = 5'$, Eq. (1.8) allows to calculate the space mean speed (here simply denoted with v). More precisely, by means of data in Table 2.4 it follows:

$$v = \frac{n}{\sum_{i=1}^{n} \frac{1}{v_i}} = \frac{12}{0.871} = 13.8\,\text{m/s} = 49.7\ \text{km/h}$$

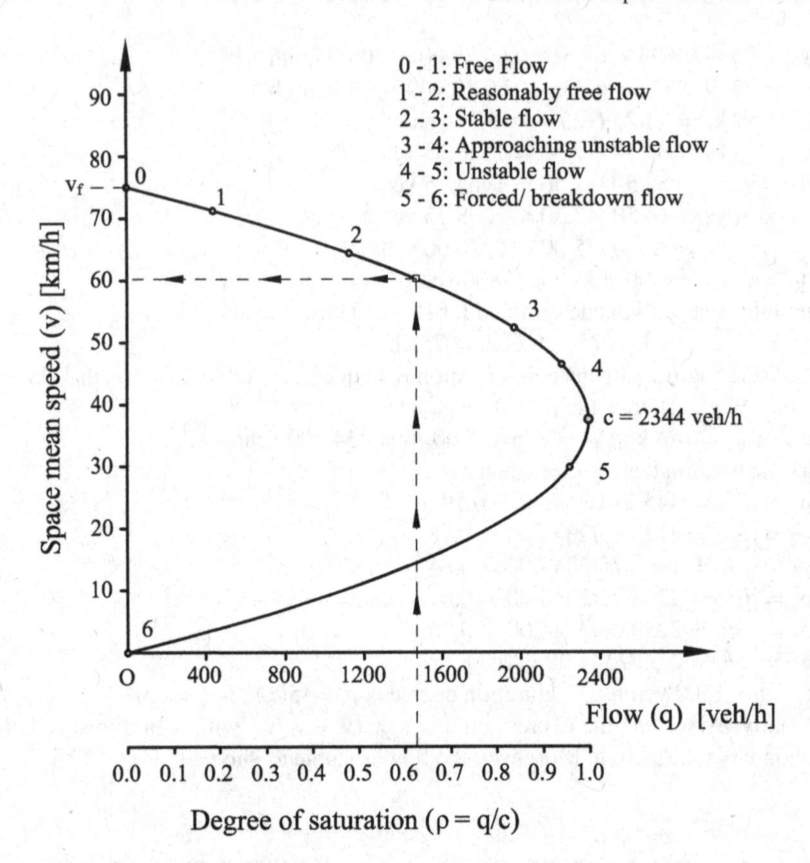

Fig. 2.12 Greenshields' relation v = v(q) for the case under study

Table 2.4 Traffic data for the first 5-min interval

Time ($\Delta T = 5'$)	Veh n°	v_i (m/s)	$1/v_i$ (s/m)
07:00:05	1	9.2	0.109
07:00:08	2	8.1	0.123
07:00:58	3	12.3	0.081
07:01:28	4	15.4	0.065
07:02:27	5	13.2	0.076
07:02:55	6	12.4	0.081
07:03:15	7	17.7	0.056
07:03:46	8	19.3	0.052
07:04:02	9	22.1	0.045
07:04:16	10	17.4	0.057
07:04:47	11	14.2	0.070
07:04:53	12	18.3	0.055
			$\Sigma = 0.871$

Table 2.5 Traffic data and assessed values for space, flow and density

Time ($\Delta T = 5'$)	q (veh/5′)	q (veh/h)	v (km/h)	k (veh/km)
07:00–07:05	12	144	49.7	2.9
07:05–07:10	110	1320	26.3	50.2
07:10–07:15	100	1200	31.4	38.2
07:15–07:20	98	1176	26.8	43.9
07:20–07:25	112	1344	33.1	40.6
07:25–07:30	114	1368	41.7	32.8
07:30–07:35	114	1368	30.2	45.3
07:35–07:40	108	1296	40.4	32.1
07:40–07:45	118	1416	39.7	35.7
07:45–07:50	80	960	11.89	80.7
07:50–07:55	50	600	6.5	92.3
07:55–08:00	56	672	8.7	77.2
08:00–08:05	112	1344	31.4	42.8
08:05–08:10	105	1260	37.56	33.5
08:10–08:15	107	1284	18.7	68.7
08:15–08:20	25	300	54.1	5.5
08:20–08:25	35	420	48	8.8
08:25–08:30	29	348	53	6.6

On the other hand, the flow is equal to $q = n/T = 12{\cdot}60'/5' = 144$ veh/h. From the fundamental flow relation (Eq. 1.32) it derives that $k = q/v = 144/49.7 = 2.9$ veh/km.

The above mentioned procedure led to the data shown in Table 2.5 referring to the intervals $\Delta T = 5'$ within an observation period $T = 1.5$ h (7:00–8:30).

Greenshields' relation (see Table 2.2) is as follows:

$$v = v_f - \frac{v_f}{k_{jam}} \cdot k$$

The parameter estimation of the model v_f (free flow speed) and k_{jam} (jam density) is carried out at least squares.

Through the 18 pairs (v; k) of Table 2.5 the regression line ($v = v(k)$) can be determined and expressed by the following equation (see Fig. 2.13a):

$$v = 55.142 - 0.5468\,k$$

Thus, it follows $v_f = 55.1$ km/h and $k_{jam} = 100.8$ veh/km (being $v_f/k_{jam} = 0.5468$). In short, the sought-after flow law is $q = 55.1{\cdot}k - 0.5468{\cdot}k^2$. The latter is shown in Fig. 2.13b together with experimental data.

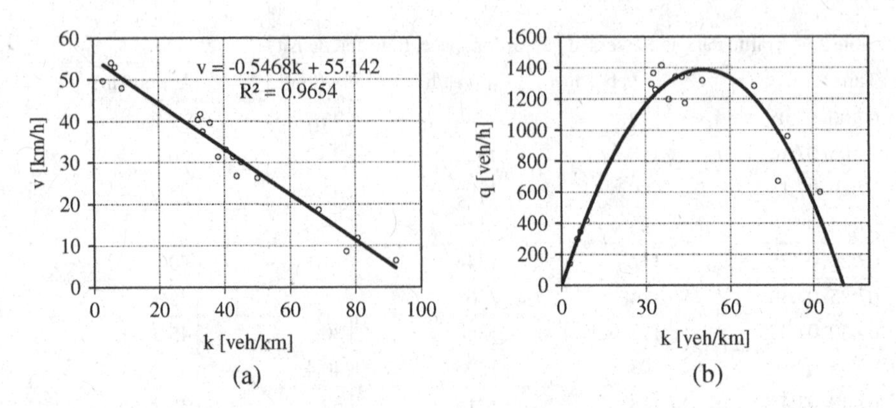

Fig. 2.13 **a** experimental pairs (v; k) and regression line; **b** calibrated Greenshields model

2.11 Case Study: Calibration of May's Flow Model

In a two-lane motorway the overtaking lane (left lane) is closed to traffic for maintenance works. On the right lane there were sampled, through inductive loops, the pairs (v; k), (q; k), (v; q), each calculated in 15-min intervals. The scatter plot of (v; k) and (v; q) are shown in Fig. 2.14. The objective is to calibrate May's bell-shaped model on the obtained data (Table 2.2) and calculate both lane capacity and limits of service levels.

May's relation has the following expression (see Table 2.2):

$$v = v_f \cdot \exp\left[-0.5 \cdot \left(\frac{k}{k_c}\right)^2\right]$$

which can be linearly expressed by means of the logarithmic transformation:

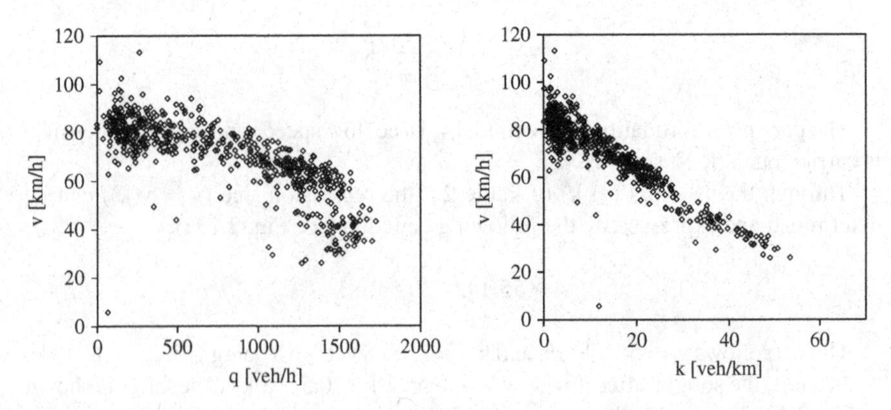

Fig. 2.14 Experimental data (v; q) and (v; k) on the lane

$$\ln(v) = \ln(v_f) - \frac{1}{2k_c^2} \cdot k^2$$

or:

$$v_1 = a + b \cdot k_1$$

where $v_1 = \ln(v)$; $a = \ln(v_f)$; $b = -1/(2k_c^2)$; $k_1 = k^2$, and where v_f denotes the free-flow speed and k_c the critical density. Through $q = k \cdot v$ the following other relations can be obtained:

$$q = v \cdot \sqrt{\frac{\ln \frac{v_f}{v}}{\frac{0,5}{k_c^2}}}$$

$$q = v_f \cdot k \cdot \exp[-(0.5/k_c^2) \cdot k^2]$$

which allow the plotting of the two further flow relations $v = v(q)$ and $q = q(k)$ to be plotted.

All this said, for each speed value v the natural logarithm ln(v) is calculated in order to infer (k; ln(v)) from each available pair (k; v).

On the basis of the relevant scatter of points (k; ln(v)), the regression line is then estimated with the least squares method and, thus, the model parameters, that is, v_f and k_c can be determined.

The equation of the regression line in Fig. 2.15 is $\ln(v) = -0.0005\ k + 4.3675$. Since $a = \ln(v_f) = 4.3675$ and $b = -1/(2k_c^2) = -0.0005$, it follows $v_f = 79$ km/h and $k_c = 32$ veh/km. Therefore, the sought-after expressions are:

$$v = 79 \cdot \exp\left[-0.5 \cdot \left(\frac{k}{32}\right)^2\right]$$

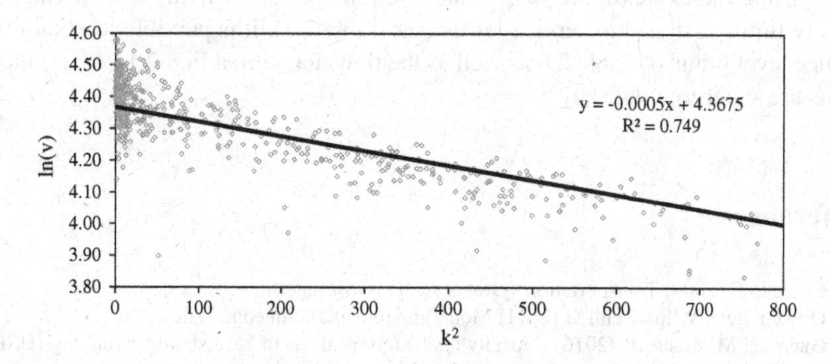

Fig. 2.15 Data from scattered points (v; k) in the graph of Fig. 2.14

Table 2.6 Limits of Levels of Service (LOS)

Flow parameters	Level of Service (LOS)				
	A	B	C	D	E
Highest density k (veh/km)	7	11	16	22	28
Lowest speed (km/h)	77	74	70	62	54
Max (q/c)	0.35	0.53	0.73	0.90	0.99
Maximum service flow q (veh/h)	540	819	1115	1372	1508

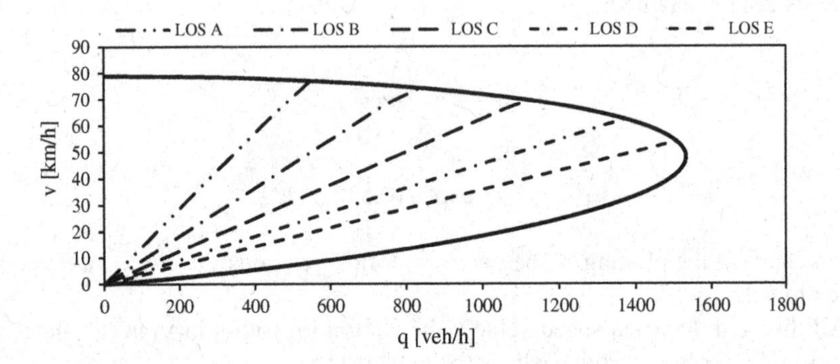

Fig. 2.16 Speed-flow curve with representation of LOS limits

$$q = 79 \cdot k \cdot \exp\left[-0.5 \cdot \left(\frac{k}{32}\right)^2\right]$$

$$q = v \cdot \sqrt{\frac{\ln \frac{79}{v}}{\frac{0.5}{32^2}}}$$

By inserting the value $k = k_c = 32$ veh/km into the previous relation $q = q(k)$, the lane capacity is obtained as follows: $c = q_{max} = 1533$ veh/h.

By using the expression $q = q(v)$ and the relation $q = k \cdot v$, and considering the density limits assigned to service levels (see Table 2.3), it is possible to obtain the service level limits of Table 2.6 as well as the flow diagram of Fig. 2.16 where these limits are superimposed.

References

1. Cascetta E (2009) Transportation systems analysis. Springer
2. Ortúzar JDD, Willumsen LG (2011) Modelling transport, 4th edn.. Wiley
3. Guerrieri M, Mauro R (2016) Capacity and safety analysis of hard-shoulder running (HSR). Motorway Case StudyTransp Res Part 92:162–183
4. Greenshields BD (1935) A study of traffic capacity. In: Highway research board proceedings

5. McShane WR, Roger OR (1990) Traffic engineering. Prentice Hall
6. Brilon W, Lohoff J (2011) Speed-flow Models for Freeways. Procedia Soc Behav Sci 16:26–36
7. Yu C, Zhang J, Yao D, Zhang R, Jin H (2016) Speed-density model of interrupted traffic flow based on coil data. Mobile Information Systems, V 2016, Article ID 7968108
8. Highway Capacity Manual, HCM (2016) Transportation research board, Washington
9. Treiber M, Kesting A (2013) Traffic flow dynamics. Springer
10. Edie LC (1961) Car following and steady-state theory for noncongested traffic. Oper Res 9:66–76
11. Koshi M, Iwasaki, M, Okhura I (1983) Some findings and an overview on vehicular flow characteristics. In: Hurdle VF, Hauer E, Steuart GF (eds) Proceedings of the 8th international symposium on transportation and traffic flow theory, Toronto, Canada. University of Toronto Press, pp 403–426
12. Zhang HM (1999) A mathematical theory of traffic hysteresis. Transp Res B, 33B, 1–23
13. Daganzo CF (2002) A behavioral theory of multi-lane traffic flow. Part I: long homogeneous freeway sections. Transp Res Part B Methodol 36(2):131–158
14. Kerner BS (2009) The physics of traffic: empirical freeway pattern features, engineering applications, and theory. Springer

Chapter 3
Continuity Flow Equation, Kinematic Waves and Shock Waves

Abstract This chapter gives an introduction to the flow continuity equation, dynamic traffic flow models and kinematic and shock waves. Since these topics are usually difficult for beginners, the mathematical treatment is highly detailed.

The traffic models in Chap. 2 are formulated in terms of macroscopic flow variables: flow, space mean speed, density.

The same variables allow to formulate other macroscopic traffic models called dynamic models.

Dynamic traffic flow models schematise the time and space evolution of traffic state in a given road element. These models assimilate a vehicle stream to a compressible fluid. Such an analogy allows to extend some results from fluid mechanics to vehicle traffic.

Besides, macroscopic dynamic models allow analysing the perturbation propagation (kinematic and shock waves) in the flow.

As later shown, these analyses are used to solve varied application problems.

The first developed dynamic model is still used in current applications. It is the LRW (Lighthill, Whitham and Richards) model [1] developed in the 1950s.

This chapter introduces this first developed dynamic traffic flow model and some of its simple applications.

3.1 Fluid Dynamic Analogy for the Traffic Flow

Consider a lane segment L of length l. The lane axis coincides with the abscissa (x-axis), so that $x \in [0, l]$.

Along L there are no entries or exits or overtakings. A flow of vehicles moving on L is assimilated to a continuous incompressible fluid.

Thus the following considerations between the traffic flow and the fluid can be taken into account.

M. Guerrieri and R. Mauro, *A Concise Introduction to Traffic Engineering*,
Springer Tracts in Civil Engineering,
https://doi.org/10.1007/978-3-030-60723-4_3

The mass conservation law is equivalent to the vehicle conservation law [2, 3], that is:

$$\frac{\partial k}{\partial t} + \frac{\partial q}{\partial x} = 0 \tag{3.1}$$

Equation (3.1) is intuitively inferred in the exemplification at the end of this section.

In (3.1) the density $k = k(x, t)$ and the flow $q = q(x, t)$ (see Chap. 1) at instant t at the point x are derivable continuous functions (at least as C1 class) of x and t.

X and t vary in the domain $\Omega \equiv [0, l]xT \subseteq R^2$. T is the width of the observation span Θ of the stream flowing on L.

As the density and the flow, also the space mean speed $v = v(x, t)$ (Eq. 1.8) is assumed at least as class C1 in Ω.

The flow function is represented by the fundamental diagram (see Chap. 2):

$$q = q(k) \tag{3.2}$$

It is generally assumed that Eq. (3.2) is yielded by the same function along the whole L and the whole T.

The *convection relation* is represented by the fundamental flow relation (Eq. 1.32):

$$q = k \cdot v \tag{3.3}$$

Equations (3.1), (3.2) and (3.3) form the fluid dynamic analogy of the traffic flow. Through (3.3) Relation (3.1) becomes:

$$\frac{\partial k}{\partial t} + \frac{\partial q}{\partial x} = \frac{\partial k}{\partial t} + \frac{dq}{dk} \cdot \frac{\partial k}{\partial x} = 0 \tag{3.4}$$

If it is set:

$$\omega(k) = \frac{dq}{dk} \tag{3.5}$$

relation (3.4) is written:

$$\frac{\partial k}{\partial t} + \omega(k) \cdot \frac{\partial k}{\partial x} = 0 \tag{3.6}$$

Relation (3.5) is the k function providing the slope of the curve (3.3).

k ranges between 0 and k_{jam}, $k \in [0, k_{jam}]$.

k_{jam} (see Sect. 2.3) is the highest density value through which the flow and speed are equal to zero.

Relation (3.6) is a quasi-linear partial differential equation (PDE).

If along the segment L there are on-ramp and off-ramp volumes, Relation (3.1) becomes:

$$\frac{\partial k}{\partial t} + \frac{\partial q}{\partial x} = r - s \qquad (3.7)$$

whrere r and s are the on-ramp and off-ramp flows, respectively, on the segment L. r and s are expressed as flow per length unit (e.g. veh/h/km). Later on in this chapter Eq. (3.5) will be examined in detail.

Equation (3.6) is briefly examined in Sect. 3.1.2.

3.1.1 Deduction of the Continuity Equation

Should neither entries nor exits (e.g. highway ramps) be between road cross sections x and x + Δx, the general flow conservation equation deduced below is to be considered.

Be n_1 and n_2 the vehicles crossing section x and section x + Δx, respectively, in the time interval [t; t + ΔT]; the flows in the two sections are $q_1 = n_1/\Delta t$ and $q_2 = n_2/\Delta t$ (Fig. 3.1).

During the time interval ΔT a variation in vehicle number is observed on the road segment Δx:

$$\Delta n = n_2 - n_1 = (q_2 - q_1) \cdot \Delta T = \Delta q \cdot \Delta T$$

Should k_1 indicate the density at instant t and k_2 the density at instant t + ΔT, the vehicle number on the segment Δx between the two time instants is $m_1 = k_1 \cdot \Delta x$ and $m_2 = k_2 \cdot \Delta x$. Therefore the vehicle variation due to the density variation is equal to:

$$\Delta m = m_1 - m_2 = (k_1 - k_2) \cdot \Delta x = -\Delta k \cdot \Delta x$$

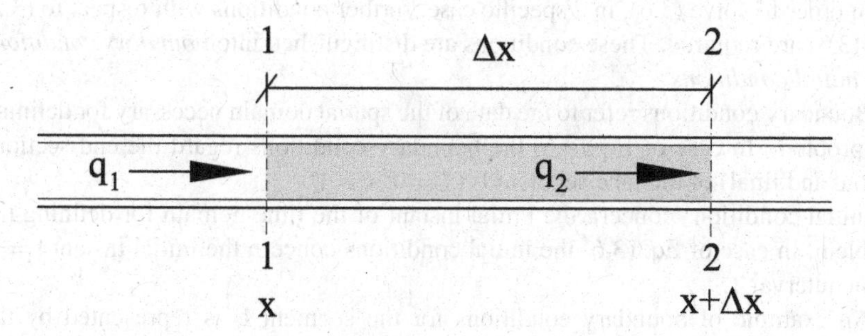

Fig. 3.1 Lane crossed by a steady-state stream

Given that by hypothesis neither entries nor exits are on the road segment under study, the condition $\Delta n = \Delta m$ must be met, and thus:

$$\Delta q \cdot \Delta T = -\Delta k \cdot \Delta x$$

or:

$$\Delta q \cdot \Delta T + \Delta k \cdot \Delta x = 0$$

By dividing by $\Delta T \cdot \Delta x$ it results that:

$$\frac{\Delta q}{\Delta x} + \frac{\Delta k}{\Delta t} = 0$$

By assimilating the road traffic to a continuous fluid on the basis of the hydrodynamic analogy (Sect. 3.1), and by letting ΔT and Δx tend to zero ($\Delta T \rightarrow 0$; $\Delta x \rightarrow 0$), it results Eq. (3.1) [1, 2]:

$$\frac{\partial q(x, t)}{\partial x} + \frac{\partial k(x, t)}{\partial t} = 0 \tag{3.1}$$

which expresses the traffic conservation law.

3.1.2 Boundary and Initial Conditions

Generally speaking, solving Eq. (3.6) means determining the density evolution over time ($t > 0$) in the cross-sections of the lane segment L.

Once the solution of (3.6)—that is, a function $k = k(x, t)$ satisfying (3.6)—is known, $q = q(x, t)$ and $v = v(x, t)$ on Ω are obtained.

In fact, $q = q(x, t)$ is yielded by (3.2) on the basis of $k = k(x, t)$.

Relation (3.3) then gives the function $v = v(x, t)$.

In order to solve (3.6), in a specific case, further conditions with respect to (3.2) and (3.3) are required. These conditions are distinguished into *boundary conditions* and *initial conditions*.

Boundary conditions refer to the data of the spatial domain necessary for defining the problem. In case of Eq. (3.6) the boundary conditions regard the end sections (initial and final) of the lane segment L ($x = 0$; $x = l$).

Initial conditions concern the initial instant of the time domain for defining the problem. In case of Eq. (3.6) the initial conditions concern the initial instant $t = t_0$ of the interval T.

An example of boundary conditions for the segment L is represented by the entering flow during the time T: $q_0(t) = q(x = 0, t)$, $t \in T$.

When the flow effects in L are studied, $q_0(t)$ is assigned for an oscillation in inflow (traffic demand).

Also the bond $q_0(t) < c_0$, $t \in T$ is a boundary condition. c_0 is the capacity of the initial section $(x = 0)$ of the segment L.

A further example of boundary conditions is given by restrictions on the off-ramp flow $q(t) = q_l(x = l, t)$ during T.

When, for instance, the effects on traffic flow due to capacity reduction at the end of L $(x = l)$ are studied, q_l is assigned.

In this case, the above boundary condition is associated to the other type $q_l(t) = c_l(t)$. c_l is the capacity of the end section of L $(x = l)$.

An example of initial conditions is given from:

$$k(x) = k(x, t = 0) = g(x) \quad x \in [0, l] \tag{3.8}$$

where g(x) is an assigned function of class C1.

Relation (3.8) continuously assigns the density k to each point x of L at time t = 0 of T. If associated to (3.6), it allows studying the density perturbations in a traffic flow (see Sect. 3.1.2).

For PDEs and therefore for (3.6) the definition of appropriate boundary conditions is also useful for "well set" problem formulations.

In PDEs a problem is "well-set" (according to Hadamard) if it has only one solution. Moreover such a solution depends continuously on the problem data. Such a change indicates that any small variation in data brings about small variations in the solution. In other words, the solution is stable with respect to small variations in input data.

In the following section a specific "well-set" problem is properly formulated for solving Eq. (3.6). It is on that problem that the calculation of kinematic waves in traffic flows quite depends.

3.1.3 Kinematic Waves

Equation (3.6) cannot be solved directly but numerically or with the *characteristics method*.

Actually, Eq. (3.6) can be solved directly in the important case of perturbations in a stationary state traffic flow (Sect. 1.2.7) along a lane of infinite length $(x \in R)$.

Therefore, for a density k(x, t) which is the result of a perturbation $\varepsilon \cdot k_1(x, t)$, it is set:

$$k(x, t) = k_0 + \varepsilon \cdot k_1(x, t) \quad x \in R, t > 0 \tag{3.9}$$

where:

$$|\varepsilon \cdot k_1(x, t)| << k_0 \tag{3.10}$$

In (3.9) k_0 is the constant density value, in space and time, of the stationary state (Sect. 1.2.7) which is perturbed.

On the other hand, the condition (3.10) indicates the smallness of density deviations from the aforesaid stationary state.

By choosing ε as small as desired, the deviations from k_0 result to be equally small.

So, Relation (3.10) also ensures that the perturbed flow (of density k(x, t)) is systematically very close to the initial stationary flow (of density $k(x, t) = k_0$ const.).

By means of the position (3.9), since k_0 and ε are assigned, the function k(x, t) is known if the function $k_1(x, t)$ is also known.

By means of (3.6), (3.9), (3.10) it is shown [4, 5] that $k_1 = k_1(x, t)$ can be obtained by solving the initial value problem (i.e. Cauchy's problem):

$$\frac{\partial k_1}{\partial t} + \omega_0 \cdot \frac{\partial k_1}{\partial x} = 0 \quad x \in R, t > 0 \tag{3.11}$$

$$k_1(x, t = 0) = g(x) \tag{3.12}$$

where

$$\omega_0 = \left(\frac{dq}{dk}\right)_{k=k_0} \tag{3.13}$$

ω_0 is a number providing the slope of the curve (3.3) (i.e. fundamental diagram) at point $P_0 \equiv (k_0, q(k_0))$.

Should the lane segment L be semi-infinite $(0 \leq L < +\infty)$ or finite $(0 \leq L \leq l)$, in order to have a "well-set" problem, (3.11) and (3.12) need to be provided with boundary conditions.

The case studies later shown in this chapter concern one of those situations.

Equation (3.12) assigns the density at time t = 0 of T to all the points x along the lane axis.

The specification g(x) in (3.12) makes the solution of (3.11) unique.

In fact, be:

$$k(x, t,) = f(x - \omega_0 \cdot t) \quad x \in R, t, > 0 \tag{3.14}$$

where $f(x - \omega_0 . t)$ is any function of class C1.

If it is set:

$$y = x - \omega_0 \cdot t \quad x \in R, t > 0 \tag{3.15}$$

relation (3.12) becomes:

$$k_1(x, t) = f(y) \tag{3.16}$$

whatever f(y) may be, with (3.16) and (3.11) it results that:

$$\frac{\partial f}{\partial y} \cdot \frac{\partial y}{\partial t} + \omega_0 \cdot \frac{\partial f}{\partial y} \cdot \frac{\partial y}{\partial x} = \frac{\partial f}{\partial y}(-\omega_0) + \omega_0 \cdot \frac{\partial f}{\partial y} \cdot 1 = 0 \tag{3.17}$$

Thus, Eq. (3.17) shows that the general solution of Eq. (3.11) is the set of functions of type $(x - \omega_0 \cdot t)$.

Now, since the arbitrary function $f(x - \omega_0 \cdot t)$ must satisfy the condition to limits (3.12), the result must be:

$$k_1(x, t = 0) = g(x) = f(x, t = 0) = f(x) \tag{3.18}$$

So, with $f(x) = g(x)$ the (unique) solution of Eq. (3.11) satisfying the initial conditions (3.12) is:

$$k_1(x, t) = g(x - \omega_0 \cdot t) \quad x \in R, t > 0 \tag{3.19}$$

For the density k(x, t) deriving from the perturbation $k_1(x, t)$ of the stationary state $(k = k_0)$ with (3.19) from (3.9), there follows:

$$k(x, t) = k_0 + \varepsilon \cdot g(x - \omega_0 \cdot t) \quad x \in R, t > 0 \tag{3.20}$$

Equation (3.20) represents the (unique) solution of Eq. (3.21) in case of an indefinite traffic flow perturbed in density starting from a stationary flow state.

Now we indicate with $\Delta k(x, t)$ the density variation with respect to the stationary state value k_0 observed in a point x at instant t:

$$\Delta k(x, t) = k(x, t) - k_0 \tag{3.21}$$

from (3.20) it yields:

$$\Delta k(x, t) = \varepsilon \cdot g(x - \omega_0 \cdot t) \quad x \in R, t > 0 \tag{3.22}$$

From (3.22) it follows that $\Delta k(x, t)$ in a generic point $x = x_i + \omega_0 \cdot t_i$ at time $t = 0$ is the same as in point $x = x_i$ at time $t = t_i$.

In fact, by replacing $(x = x_i + \omega_0 \cdot t_i, t = 0)$ in (3.22), it yields:

$$\varepsilon \cdot g(x_i + w_0 \cdot t_i - \omega_0 \cdot 0) = \varepsilon \cdot g(x_i + \omega_0 \cdot t_i) \tag{3.23}$$

Equation (3.23) indicates that density variations $\Delta k(x, t)$ move along the traffic flow with speed ω_0.

It is worth reminding that ω_0 (Eq. 3.13) is the slope of the curve (3.3) at point $P_0 \equiv (k_0, q(k_0))$ which identifies the considered stationary state.

The observations made so far suggest that density variations from k_0 move along the traffic flow as kinematic waves with speed ω_0.

As previously said in Chap. 2, the curve (3.3), $q = k \cdot v$, has generally a maximum density value $k = k_c$.

k_c subdivides the interval $[0, k_{jam}]$ into two parts (see Sect. 2.3). When $0 \leq k < k_c$, ω_0 is positive. When $k_c \leq k < k_{jam}$, ω_0 is negative.

With a positive ω_0, any density variation (or density perturbation $\varepsilon \cdot k_1(x, t)$) triggered in a point, propagates forward along the traffic flow.

With a negative ω_0 it propagates backwards.

What said so far, becomes much more evident thanks to the following considerations.

From (3.32) and (3.21) it follows:

$$g(x - \omega_0 \cdot t) = [\Delta k(x, t)/\varepsilon] \tag{3.24}$$

Denoting the inverse function of $g(\cdot)$ with $g^{-1}(\cdot)$, from (3.24) it follows:

$$x - \omega_0 \cdot t = g^{-1}(\Delta k/\varepsilon) \tag{3.25}$$

If $\Delta k/\varepsilon$ is constant, Eq. (3.25) is the equation of a straight line on the plane (t, x). In other words, the straight line (3.25) is the locus of points on the plane (t, x) in which the density variations Δk is constant.

Figure 3.2 clarifies what said so far. The half-line (B-A) refers to $\omega_0 > 0$. It is immediately clear that the perturbation Δk in point x at instant t (point A) is equal to that observed at instant t_1 ($t_1 < t$, that is, before) in point $x_1 = x - \omega_0 \cdot (t - t_1)$ (point B). The half-line (C-D) refers to $\omega_0 < 0$. In this case Δk at a point x' at instant

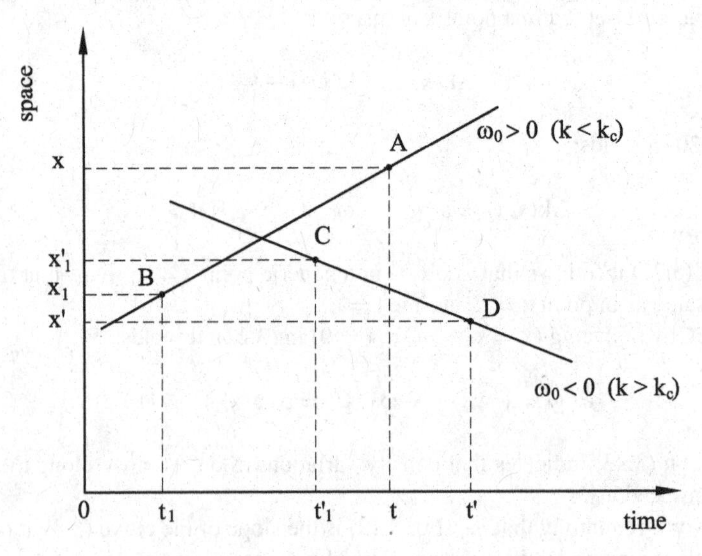

Fig. 3.2 Density perturbations in space and time for $\omega_0 > 0$ and for $\omega_0 < 0$

t' (point D) is equal to Δk observed at instant t'_1 ($t'_1 < t'$, that is, before) at point $x'_1 = x' + \omega_0 \cdot (t' - t'_1)$.

In short, if $\omega_0 > 0$ (i.e. $k < k_c$) the perturbation propagates forward from B to A ($t > t_1 \rightarrow x > x_1$).

If $\omega_0 < 0$ (i.e. $k \geq k_c$) the perturbation propagates backwards from C to D (for $t' > t \rightarrow x' < x_c$).

The half-lines in Fig. 3.2 are part of lines termed "*characteristics*" in the PDE theory [4, 6].

The characteristic lines ($x = \text{cost} + \omega_0 \cdot t$) on the plane ($t > 0$, R) are the locus of the points where not only $k(x, t)$ is constant, but $q = q(x, t)$ and $v = v(x, t)$ are also constant for (3.2) and (3.3).

Thus, if an observator moves parallel to the road with speed $\omega_0 = (dq/dk)_{k=k_0}$ (see (3.13)), he/she would not identify any flow variation $q = q(x, t)$.

The speed ω_A of a kinematic wave characterised by flow q_A and density k_A, as already said, is given by the slope of the tangent to the curve $q = q(k)$ at point A, whose coordinates are: $A \equiv (k_A, q_A)$.

The speed of the flow v_A, on the other hand, is measured on the curve $q = q(k)$ from the slope of the secant OA (see Fig. 2.5).

There is always $v_A > \omega_A$ whatever the position A may be in the curve $q = q(k)$.

The speed of the kinematic waves for $k < k_c$, and for $k > k_c$ is thus lower than that of the traffic flow.

The characteristic lines have other numerous mathematical properties. They are, moreover, an important calculation tool for solving some types of PDEs. This topic is, however, beyond the scope of this book. So refer, among the others, to [4] for the characteristic lines.

3.1.4 *Perturbations of the Traffic Flow and Speed*

Solving Eq. (3.11) also allows to calculate the space and time evolution of perturbations of $q = q(x, t)$ and $v = v(x, t)$ resulting from a density variation Δk.

In fact, with:

$$k(x, t) = k_0 + \varepsilon \cdot g(x - \omega_0 \cdot t) \quad x \in R, t > 0 \tag{3.26}$$

by applying (3.2) to each of the two terms of (3.26):

$$q(k(x, t)) = q(k_0) + \varepsilon \cdot q(g(x - \omega_0 \cdot t)) \tag{3.27}$$

or:

$$q(x, t) = q_0 + \varepsilon \cdot h(x - \omega_0 \cdot t) \tag{3.28}$$

where the function $h(x - \omega_0 \cdot t)$ depends on the functional form of (3.2) and of $g(\cdot)$.

With (3.4) and (3.5) it follows

$$\frac{q(x,t)}{k(x,t)} = \frac{q_0}{v_0} + \varepsilon \cdot \frac{h(x - \omega_0 \cdot t)}{g(x - \omega_0 \cdot t)} \tag{3.29}$$

By recovering (3.3) it yields:

$$v(x,t) = v_0 + \varepsilon \cdot d(x - \omega_0 \cdot t) \tag{3.30}$$

The function $d(x - \omega_0 \cdot t)$ depends on the functional form of $h(x - \omega_0 \cdot t)$ and $g(x - \omega_0 \cdot t)$.

Equations (3.28) and (3.30) thus provide the desired expressions of the traffic flow and speed starting from the stationary state ($q = q_0$, $v = v_0$ and $k = k_0$) as a result of a small perturbation $\varepsilon \cdot g(x - \omega_0 \cdot t)$ of k_0.

3.2 The LWR Model and Shock Waves

When the traffic flow law (3.2) is the Greenshields diagram (see Sect. 2.3.1), the model resulting from Eqs. (3.1), (3.2) and (3.3) is called LWR model.

LWR is the acronym from the surnames of the researchers who developed the model. They are Lighthill, Whitham and Richards [1].

Thus, from the expression in Table 2.2 for $q = q(k)$ (Fig. 3.3), it follows:

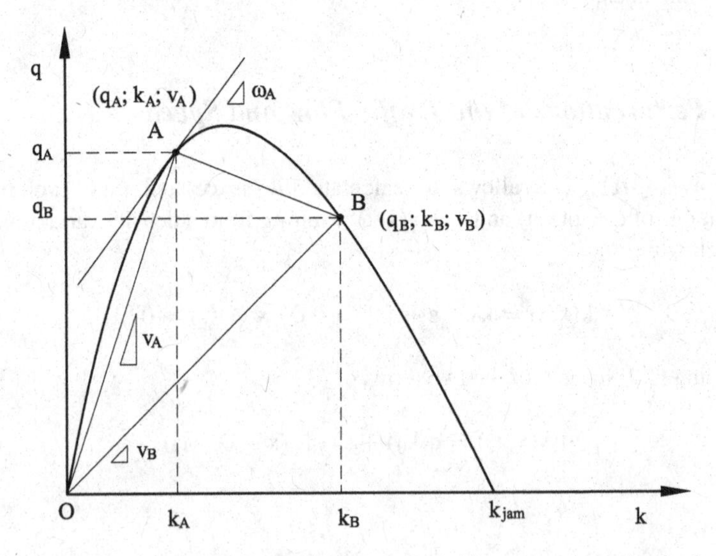

Fig. 3.3 Fundamental diagram of Greenshields; traces of a kinematic wave and a shock wave

$$q = v_f \cdot \left(k - \frac{k^2}{k_{jam}}\right) x \in [0, k_{jam}] \tag{3.31}$$

With (3.31) for the speed ω_0 of a kinematic wave, it yields:

$$\omega_0 = \left(\frac{dq}{dk}\right)_{k=k_0} = v_f \cdot \left(1 - \frac{2k_0}{k_{jam}}\right) \tag{3.32}$$

In Fig. (3.3) point A represents a flow stationary state characterised by $k = k_A$ and $q = q_A$.

The speed corresponding to k_A and q_A for (3.3) is measured by the slope of the secant OA.

On the other hand, the slope of the tangent in A to the curve (3.31) in Fig. 3.3 measures the speed ω_A on the basis of (3.32).

ω_A is the kinematic wave speed transporting a small perturbation of the stationary state A along the traffic flow $\equiv (k_A, q_A, v_A)$.

In case of Fig. 3.3, since $\omega_A > 0$, the wave propagates in the same direction as the traffic flow. Moreover, since $\omega_A < v_A$, the wave moves more slowly than the vehicles of this flow.

Now, at a certain instant, a modification of the flow state is hypothesised from A $\equiv (k_A, q_A, v_A)$ to B $\equiv (k_B, q_B, v_B)$.

In Fig. 3.3 state B is characterised by a speed $v_B < v_A$. This occurs, for instance, if a platoon of vehicles A moving with a speed v_A reaches a slower platoon B, running with a speed $v_B < v_A$.

In the road section where the two platoons merge, there is discontinuity in the flow state. The discontinuity in the flow regime is called *shock wave*.

Figure 3.4 illustrates this situation.

Moreover, Fig. 3.4 shows that the shock wave between A and B is inclined to the axis t of the same value as the chord AB on the axis k.

This inclination to the plane (t, x) is evidently the speed value w_s of the shockwave.

Actually, w_s is obtained by simple geometric considerations (Fig. 3.3):

$$w_s = \frac{q_B - q_A}{k_B - k_A} = \frac{q_A - q_B}{k_A - k_B} \tag{3.33}$$

The physical meaning of (3.3) can be obtained from the following line of reasoning.

On one side of the discontinuity S, the flow state A defined by k_A, q_A, v_A; on the other side of S, the flow state B is defined by k_B, q_B, v_B.

This situation is schematised in Fig. 3.5.

Now consider an observer travelling at speed w_s along the lane axis.

This observer sees state A vehicles as moving at speed $v_A - w_s$.

State B vehicles appear to move at speed $v_B - w_s$.

For an observer in motion during an interval Δt, the flow $q_A = N_A/\Delta t$ coming out from S, by means of (3.3), is as such:

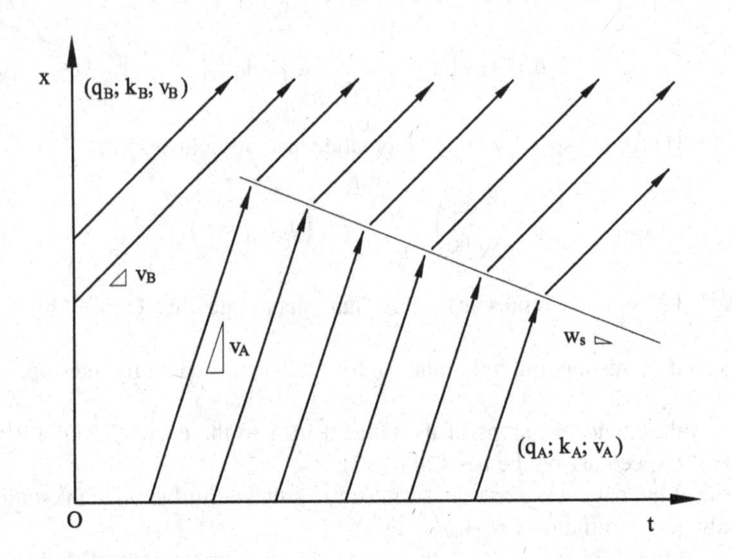

Fig. 3.4 Discontinuity in vehicle trajectories and shock waves

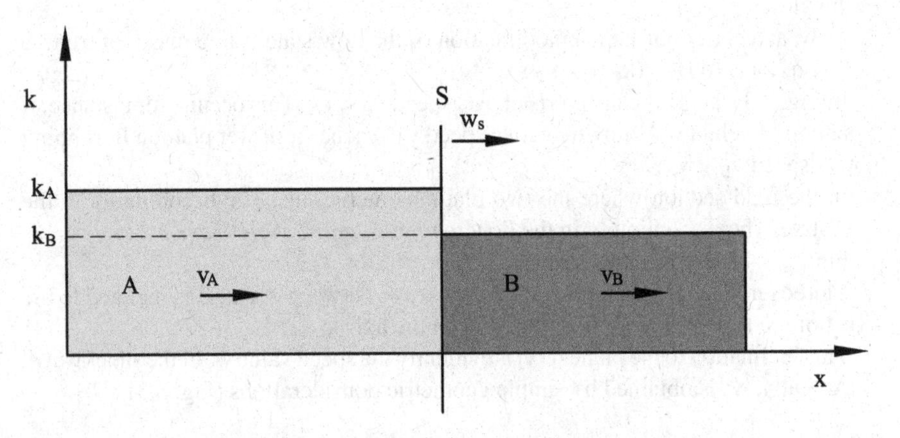

Fig. 3.5 Schematic representation of the frontal shock wave section S

$$q_A = \frac{N_A}{\Delta t} = (v_A - w_s) \cdot k_A \tag{3.34}$$

The same flow coming ot from S is assessed by an observer in the motion equal to:

$$q_A = q_B = \frac{N_B}{\Delta t} = (v_B - w_s) \cdot k_B \tag{3.35}$$

With (3.35) and (3.34) we obtain:

$$k_A \cdot (v_A - w_s) \cdot \Delta t = k_B \cdot (v_B - w_s) \cdot \Delta t \tag{3.36}$$

Expression (3.36) illustrates the *conservation-of-vehicles* principle in case of discontinuity of the flow regime in finite terms.

In fact, in S there is a variation in density and speed, but not in vehicle quantity, since there are neither entries nor exits upwards and downwards of S.

From (3.36) it immediately follows that:

$$w_s = \frac{v_B k_B - v_A k_A}{k_B - k_A} \tag{3.37}$$

Expression (3.37) for (3.35) is written as:

$$w_s = \frac{q_B - q_A}{k_B - k_A} \tag{3.38}$$

which coincides with (3.33).

From (3.31), for (3.3) it yields:

$$v = v_f \cdot \left(1 - \frac{k}{k_{jam}}\right) \quad x \in [0, k_{jam}] \tag{3.39}$$

Equation (3.39) is the relation $v = v(k)$ of Greenshields (see Table 2.2).

With Eq. (3.39) and Eq. (3.37) w_s is written as:

$$w_s = \frac{\left[k_B \cdot v_f \cdot \left(1 - \frac{k_B}{k_{jam}}\right) - k_A \cdot v_f \cdot \left(1 - \frac{k_A}{k_{jam}}\right)\right]}{k_B - k_A} \tag{3.40}$$

If we set $\eta_A = k_A/k_{jam}$ and $\eta_B = k_B/k_{jam}$, namely normalizing k_A and k_B with k_{jam} Eq. (3.40) becomes:

$$w_s = \frac{[k_B \cdot v_f \cdot (1 - \eta_B) - k_A \cdot v_f \cdot (1 - \eta_A)]}{k_B - k_A} \tag{3.41}$$

By eliminating k_A and k_B from (3.41), Eq. (3.41) is written as:

$$w_s = v_f \cdot [1 - (\eta_A + \eta_B)] \tag{3.42}$$

Equation (3.42) can be exemplified for different traffic situations in uninterrupted flow conditions.

Some of these typical situations will be examined in the sections with case studies.

3.2.1 *Case Study: Queue Formation and Dissipation Due to Presence of a Heavy Vehicle on a Two-Lane Undivided Highway*

On a two-lane highway with a single carriageway, the overtaking is forbidden [4, 6]. On one of the two lanes the curves $q = q(k)$ and $v = v(k)$ were deduced. In a certain time interval a flow $q_1 = 1102$ pcu/h, a density $k_1 = 16$ pcu/km and a space mean speed $v_1 = 70$ km/h are identified. At a certain instant, a heavy vehicle coming from a side entry gets into the studied lane, starts travelling at a constant speed $v_2 = 30$ km/h, and forces the follower vehicles to keep the same speed, since the overtaking manoeuvre is forbidden. As a consequence, a vehicle platoon is formed that travels with $v_2 = 30$ km/h, density $k_2 = 45$ pcu/km and flow $q_2 = 1336$ pcu/h (Fig. 3.6), respectively inferred from the curves $v = v(k)$ and $q = q(k)$.

Between the traffic flow (q_2; k_2; v_2) and the flow after it—in other words, that one coming from upward of the perturbation which, in its turn, is induced by the slow vehicle—the resulting shock wave assumes the following speed:

$$w_{1,2} = \frac{q_1 - q_2}{k_1 - k_2} = \frac{1102 - 1336}{16 - 45} = 8\,\text{km/h}$$

Since $w_{1,2} > 0$, the shock wave propagates in the same direction as the vehicle stream and is the queue tail behind the heavy vehicle. The head queue moves with the same speed as the heavy vehicle $v_2 = 30$ km/h and therefore the queue extends with the speed $v' = 30 - 8 = 22$ km/h.

After 3 km from the side entry, the heavy vehicle exits, getting into a side road. The 3-km-long segment has been travelled in a time $\Delta t = s/v_2 = 3/30 = 0.1\text{ h} = 6'$, thus forming, behind the heavy vehicle, a long queue $L = \Delta t \cdot v' = 0.1 \cdot 22 = 2.2$ km, formed by $m = L \cdot k_2 = 2.2 \cdot 45 = 99$ vehicles.

So, vehicles can increase their own speeds but, when they get to a winding segment with a speed limit of 50 km/h, they conform their own driving to that value. It then follows a third flow state with speed $v_3 = 50$ km/h, density $k_3 = 31$ pcu/km and flow $q_3 = 1530$ pcu/h (Fig. 3.6). A new shock wave develops and divides this flow (q_3; k_3; v_3) from that one of the queuing vehicles (q_2; k_2; v_2) with the following speeds:

$$w_{2,3} = \frac{q_2 - q_3}{k_2 - k_3} = \frac{1336 - 1530}{45 - 31} = -14\,\text{km/h}$$

The shock wave represents the new head of the queue and moves in the opposite direction to the tail queue. Consequently, the queue tends to dissipate with speed equal to the sum of the head/tail speeds: $\Delta w = 8 + 14 = 22$ km/h. With respect to the exit instant of the heavy vehicle, the queue completely dissipates after a period $\Delta t^* = L/\Delta w = 2.2/22 = 0.1\text{ h} = 6'$.

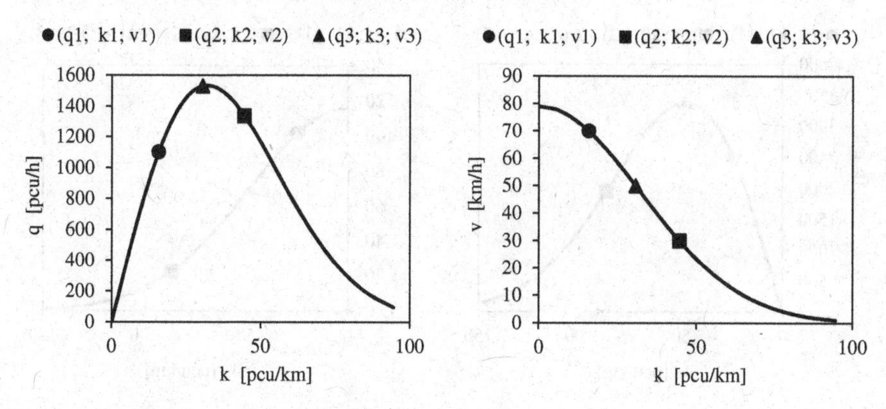

Fig. 3.6 q = q(k) and v = v(k) relationships for the lane

3.2.2 *Case Study: Estimation of the Effects of an Accident on the Flow in a Two-Lane Dual Carriageway Highway*

A highway with two-lane for each carriageway is marked by the values of the flow parameters shown in Table 2.1. More in detail, the carriageway capacity is 3370 pcu/h while the capacity of the overtaking lane is 1950 pcu/h.

One carriageway is travelled by a traffic flow of $q_1 = 2500$ pcu/h with speed $v_1 = 102$ km/h and density $k_1 = 25$ pcu/km.

At a certain time instant, due to an accident, the right lane is closed for some hundreds of metres, while the overtaking lane still remains open to transit. Upstream of the closed segment of the driving lane, there occurs a queue formation, since the initial traffic flow is greater than the capacity of the overtaking lane ($q_1 > 1950$ pcu/h). More in detail, the flow will take place with a flow equal to the capacity value of the only still open lane, thus resulting in $q_2 = 1950$ pcu/h, corresponding to a density of $k_2 = 87$ pcu/km and a speed $v_2 = 22$ km/h from the fundamental diagram $q = q(k)$ describing the whole carriageway. Between this new state flow (q_2; k_2; v_2) and that deriving from upstream (q_1; k_1; v_i) (Fig. 3.7), there occurs the formation of a shock wave with speed

$$w_{1,2} = \frac{q_1 - q_2}{k_1 - k_2} = \frac{2500 - 1950}{25 - 87} = -9\,\text{km/h}$$

As $w_{1,2} < 0$, the shock wave representing the tail queues starting at the interruption section of the right lane, propagates in opposite direction to the motion of vehicles with speed $v' = 9$ km/h. This speed brings about the increase in queue length upstream of the restriction; while vehicles beyond the restriction will merge into a flow of 1950 pcu/h equal to the capacity of the overtaking lane and an average speed equal to the critical speed $v_c = 78$ km/h (see Table 2.1).

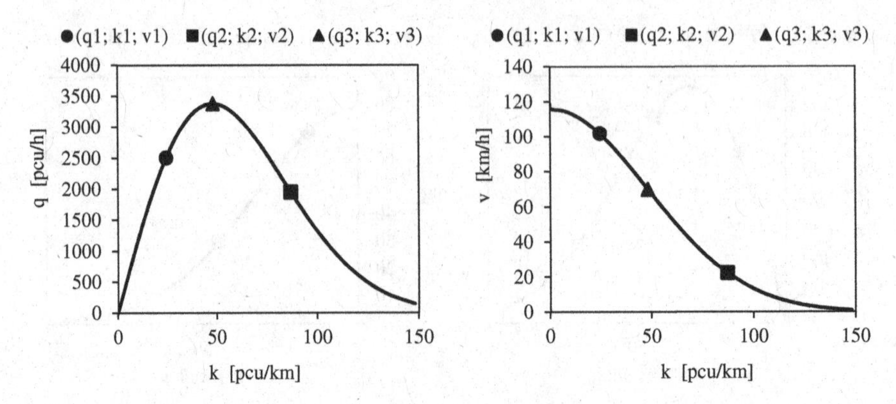

Fig. 3.7 q = q(k) and v = v(k) relationships of the carriageway

After two hours from the closing instant, the driving lane is reopened. During the closing time a long queue was formed and on the whole reaches length $L = \Delta t \cdot v' = 2 \cdot 9 = 18$ km. Starting from the reopening instant of the whole carriageway, the vehicles increase in speed, thus forming a traffic flow $q_3 = 3370$ pcu/h (equal to capacity) with speed $v_3 = 70$ km/h and density $k_3 = 48$ pcu/km (Fig. 3.7).

Thus, a new shockwave develops and divides, instant after instant, this new flow from that formed by vehicles still in queue.

$$w_{2,3} = \frac{q_2 - q_3}{k_2 - k_3} = \frac{1950 - 3370}{87 - 48} = -36\,\text{km/h}$$

The shockwave represents the new head of the queue which, being negative in value, moves in the opposite direction to the tail of the queue. Although the two fronts move in the same direction, being $w_{2,3} > w_{1,2}$ the queue dissipates with speed $\Delta w = w_{2,3} - w_{1,2} = 36 - 9 = 27$ km/h, in a period of time $\Delta t^* = L/\Delta w = 18/27 = 0.667$ h = 40 min.

References

1. Lighthill M, Whitham G (1955) On kinematic waves II. A theory of traffic flow on long crowded roads. Proc R Soc Lond Part A 229 (1178):317–345
2. Gerlough DL, Huber MJ (1975) Traffic flow theory; a monograph. Special report 165. TRB
3. Kühne R, Michalopoulos P, Zhang HM (1992) Continuum flow models, Chapter 5, in Traffic flow theory(http://www.tfhrc.gov)
4. Leutzbach W (1988) Introduction to the theory of traffic flow. Springer, Berlin
5. Prigogine I, Herman R (1971) Kinetic theory of vehicular traffic. Elsevier
6. Ferrari P (2007) Theory and control of traffic flow. TEP-Tipografia Editrice Pisana (in Italian)

Chapter 4
Microscopic Models and Traffic Instability

Abstract This chapter describes some of the main microscopic models (linear and nonlinear). These dynamic traffic models involve studying local instability, asymptotic instability and flow breakdown. Moreover, a genealogy of the main traffic models (macroscopic, mesoscopic and microscopic) is reviewed in a synthetic way.

4.1 Microscopic Models: Car-Following Theory

Among microscopic models, car-following models [1] are used to describe the behaviour of the driver-vehicle system in a traffic stream. Car-following models are implemented in traffic microsimulation software (Aimsun, Paramics, Vissim, MATSime, SUMO etc.), along with lane-changing [2, 3], overtaking, gap-acceptance (see Chap. 7) and traffic generation models.

Car-following models are based on a behavioural principle: a vehicle driver responds to the generic stimulus at time t, induced by the driving speed of the preceding vehicle, accelerating or decelerating from a later instant in time $t + t_{pr}$ (t_{pr} being the driver perception-reaction time) and proportional to the result of the stimulus multiplied by driver sensitivity [1]:

$$\text{Response}\left(t + t_{pr}\right) = \text{sensitivity} \cdot \text{stimulus}\ (t) \tag{4.1}$$

In *car-following* models a stimulus is a function of the speed difference between two given vehicles in a platoon, while sensitivity assumes a different form according to the selected model. Therefore, these models are able to simulate the real dynamics from a phenomenological point of view starting from very simple basic assumptions.

4.2 Linear Model

Consider the scheme of Fig. 4.1 which represents two vehicles in one-lane traffic

M. Guerrieri and R. Mauro, *A Concise Introduction to Traffic Engineering*,
Springer Tracts in Civil Engineering,
https://doi.org/10.1007/978-3-030-60723-4_4

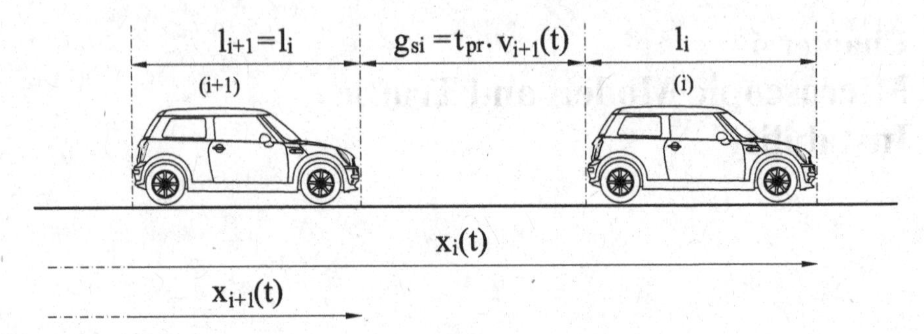

Fig. 4.1 A basic schematization of the car-following model

stream. Be "i" the lead vehicle and "i + 1" the follower vehicle. For prudential reasons the follower vehicle, at generic time instant t, must keep a net distance $g_{si}(t)$ behind the vehicle ahead (see Fig. 4.1) that has a value proportional to its travel speed ($v_{i+1}(t)$), or:

$$g_{si}(t) = t_{pr} \cdot v_{i+1}(t) \tag{4.2}$$

where t_{pr} is the perception-reaction time (at a constant value only prima facie). Thus, the space headway between the two vehicles is:

$$x_i(t) - x_{i+1}(t) = t_{pr} \cdot v_{i+1}(t) + l_i \tag{4.3}$$

By deriving the previous equation with respect to time and bearing in mind that the follower vehicle's driver response in terms of acceleration is delayed in tpr time (i.e. perception-reaction time) with regard to the time instant of the stimulus, it follows:

$$a_{i+1}(t + t_{pr}) = \frac{1}{t_{pr}} \cdot [v_i(t) - v_{i+1}(t)] \tag{4.4}$$

Empirical experiences show that in order to better approximate the real user behaviours, it is more advisable to use a coefficient λ in (4.4) that has a value not coincident with the reciprocal of the perception-reaction time ($\lambda \neq 1/t_{pr}$). The coefficient λ represents the driver sensitivity introduced in Eq. (4.1). Thus, the stimulus–response equation of car-following model can be rewritten as:

$$a_{i+1}(t + t_{pr}) = \lambda \cdot [v_i(t) - v_{i+1}(t)] \tag{4.5}$$

In short, given two vehicles in a platoon, the (positive or negative) acceleration made by the follower vehicle is obtained from the difference between the spot speeds of the two vehicles multiplied by the driver sensitivity coefficient λ which, for known boundary conditions, can be assumed as constant.

4.3 Impulsive Variations of Vehicle Speeds: Traffic Instability and Stop ('phantom Traffic Jams')

Consider an urban road with the traffic lights (see Chap. 10). During the red phase only two vehicles are stopped at the traffic lights, with a 10-m distance between each other. As soon as the green phase is displayed, the lead vehicle starts moving and instantaneously reaches the speed of 50 km/h which is then maintained constant over time. After this increase in speed, the follower vehicle starts moving in conformity with Eq. (4.5). Suppose that $\lambda = 1\ s^{-1}$; $t_{pr} = 1$ s and that the laws of motion for the two vehicles be determined by discrete intervals $\Delta t = 1$ s (assuming, for the sake of simplicity, that at each interval Δt the acceleration of the vehicle in queue is constant and equal to the average of the values calculated at the end of each interval). Equation (4.5) makes it possible to obtain for the follower vehicle (vehicle 2) both the speed equation ($v_2 = v_2(t)$) and the space travelled as a function of time ($x_2 = x_2(t)$):

$$v_2(t) = v_2(t - \Delta t) + \frac{1}{2}[a_2(t - \Delta t) + a_2(t)] \cdot \Delta t \tag{4.6}$$

$$x_2(t) = x_2(t - \Delta t) + \frac{1}{2}[v_2(t - \Delta t) + v_2(t)] \cdot \Delta t \tag{4.7}$$

The numerical results are shown in Table 4.1. In Figs. 4.2 and 4.3 the diagrams illustrate the speed changes (v_1-v_2) and the spacing between the two vehicles as a function of time, respectively. It is worth observing that the values of ($v_1 - v_2$) and ($x_1 - x_2$) have an oscillatory behaviour and stabilise from 16th second.

Figures 4.4 and 4.5 refer to a case study similar to the previous one but with more numerous vehicles in a platoon.

For simplicity, the focus here is on the first four vehicles of the platoon, distanced each other by 10 m at the initial instant.

Visibly in this instance, the impulsive variation in speed of the leader vehicle (which instantaneously moves from 0 to 50 km/h) leads to increasing amplitude in oscillation of the vehicles speeds in a platoon (which are calculated with the procedure described above).

These oscillations entail a progressive speed reduction in the farthest vehicles of a platoon up to any temporary traffic stops (in the case under study vehicle 4 settles down at instant $t = 9$ s, see Fig. 4.4) with possible queue formation.

An analogous result is obtained when considering the effect of impulsive speed variation on a motorway (uninterrupted flow).

Take, for instance, the case of 5 vehicles in a platoon, each other distanced by 20 m and travelling on a motorway lane at speeds $v_1 = v_2 = v_3 = v_4 = v_5 = 100$ km/h (assuming $\lambda = 1\ s^{-1}$; $t_{pr} = 1$), as represented in Fig. 4.5. If at a given time instant the lead vehicle (vehicle 1) slows down from 100 to 80 km/h (for instance, consequent on a vehicle entering from a on-ramp), it results in a speed decrease in the entire platoon which settles down at 80 km/h. However, there is a transient state in which

Table 4.1 Values calculated for accelerations, speeds and distances over time

Time (s)	v_1 (m/s)	a_2 (m/s^2)	v_2 (m/s)	v_1–v_2 (m/s)	x_1 (m)	x_2 (m)	x_1–x_2 (m)
0	13.9	0.0	0.0	13.9	0.0	– 10.0	10.0
1	13.9	13.9	0.0	13.9	13.9	– 10.0	23.9
2	13.9	13.9	13.9	0.0	27.8	– 3.1	30.8
3	13.9	0.0	20.8	– 6.9	41.7	14.3	27.4
4	13.9	– 6.9	17.4	– 3.5	55.6	33.4	22.2
5	13.9	– 3.5	12.2	1.7	69.4	48.2	21.3
6	13.9	1.7	11.3	2.6	83.3	59.9	23.5
7	13.9	2.6	13.5	0.4	97.2	72.2	25.0
8	13.9	0.4	15.0	– 1.1	111.1	86.5	24.6
9	13.9	– 1.1	14.6	– 0.8	125.0	101.3	23.7
10	13.9	– 0.8	13.7	0.2	138.9	115.5	23.4
11	13.9	0.2	13.4	0.5	152.8	129.0	23.7
12	13.9	0.5	13.7	0.1	166.7	142.6	24.0
13	13.9	0.1	14.0	– 0.2	180.6	156.5	24.0
14	13.9	– 0.2	14.0	– 0.2	194.4	170.6	23.9
15	13.9	– 0.2	13.9	0.0	208.3	184.5	23.8
16	13.9	0.0	13.8	0.1	222.2	198.4	23.9
17	13.9	0.1	13.9	0.0	236.1	212.2	23.9
18	13.9	0.0	13.9	0.0	250.0	226.1	23.9
19	13.9	0.0	13.9	0.0	263.9	240.0	23.9
20	13.9	0.0	13.9	0.0	277.8	253.9	23.9

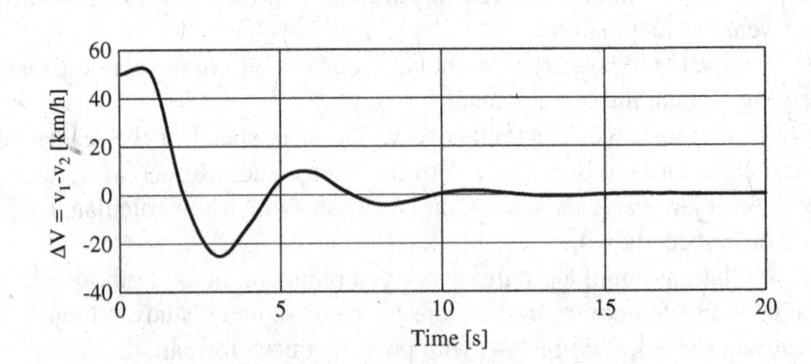

Fig. 4.2 Variation in the speed difference between the two vehicles over time

every vehicle—except the lead vehicle—has speed oscillations (v_i = variable) with more and more decreasing amplitude over time until the value becomes uniform to the lead vehicle speed (80 km/h).

As evident from Fig. 4.6, speed oscillations are smaller for next-ahead vehicles and tend to become higher for the farthest. Thus, if the platoon is quite numerous,

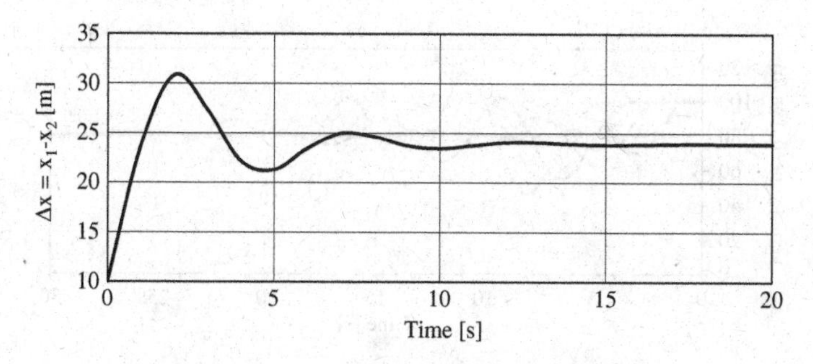

Fig. 4.3 Variation in the space headway between the two vehicles over time

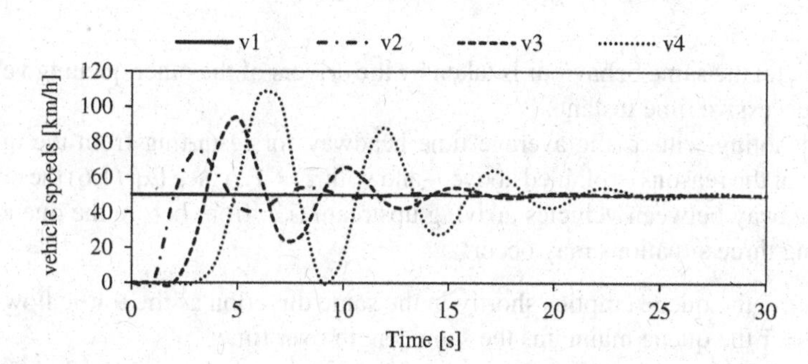

Fig. 4.4 Variation over time in speeds of 4 vehicles in a platoon following an impulsive increase in the lead vehicle speed at an intersection

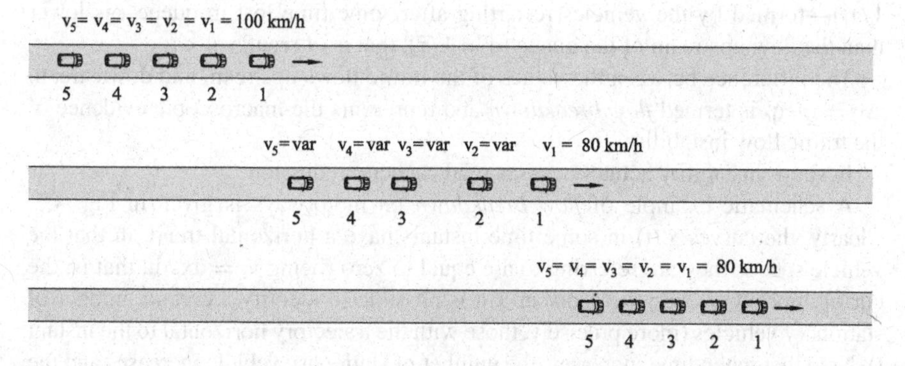

Fig. 4.5 States of a platoon movement due to the lead vehicle instantaneous deceleration

significant speed reductions will occur until one vehicle and those after it stop and lead to a queue-forming condition.

If this is the case, the vehicle first coming to rest, will move from the head of the queue when the driver considers the safety distance from the vehicle ahead to be

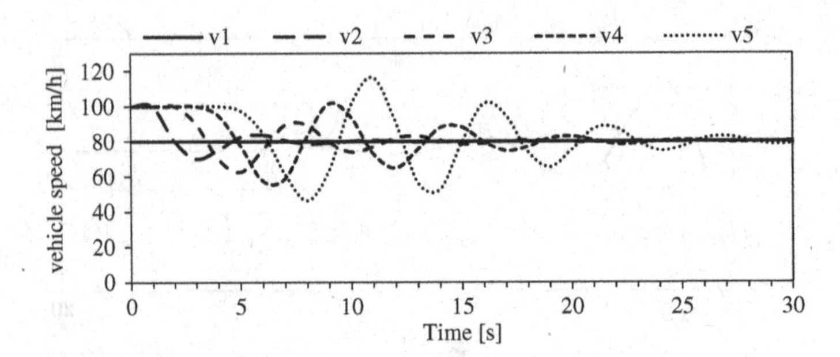

Fig. 4.6 Variation of vehicle speeds in a platoon due to an impulsive reduction in the lead vehicle speed (motorway)

appropriate; the same behaviour is taken by the drivers of the other queuing vehicles in the successive time instants.

By denoting with τ_r the average time headway for restarting from the queue - formed for the reasons explained above—and with $\overline{\tau} = 1/q$ (see Eq. (1.6) the average time headway between vehicles arriving upstream (i.e. from before the queue), the following three situations may occur:

- if $\tau_r < \overline{\tau}$ the queue empties shortly in the same direction as the traffic flow;
- if $\tau_r = \overline{\tau}$ the queue maintains the same length over time;
- if $\tau_r > \overline{\tau}$ the queue length increases over time, propagating in the opposite direction of the traffic flow (that is, backwards).

In the latter condition, since $\tau_r > \overline{\tau}$, the flow downstream from the queue ($q_r = 1/\tau_r$)—formed by the vehicles restarting after some time lost in queue—is lower than the flow upstream of the queue ($q = 1/\overline{\tau}$), that is, it results $q_r < q$.

The difference between the values of the traffic flows upstream and downstream $\Delta q = q - q_r$ is termed *flow breakdown* and represents the macroscopic evidence of the traffic flow instability.

In short, instability actually causes road capacity reduction.

A schematic example of *flow breakdown* on motorways is given in Fig. 4.7. Clearly, the curves $x_i(t)$ in some time instants have a horizontal trend, in that the vehicle speeds they are referring to are equal to zero (being $v_i = dx_i/dt$, that is, the curve slope). For every time instant t it is possible to identify a certain number of stationary vehicles (more precisely, those with the trajectory horizontal to the instant t). Note that when time increases, the number of stationary vehicle increases and the block section (value in abscissa axis) gradually decreases; thus, the queue propagates backwards to the road, moving upwards of the flow.

Even in case of no queue-formation, the impulsive speed variations can lead to dangerous conditions for users. As a rule, the distances between vehicle pairs travelling one after the other (also these with oscillatory behaviour) tend to decrease for the farthest vehicles in a platoon, with considerable risks of *rear-end chain collisions*. Notice, for instance, in Fig. 4.8 that at time instant t = 7 s the distance (x_4-x_5) between vehicle 4 and vehicle 5 is only 5 m.

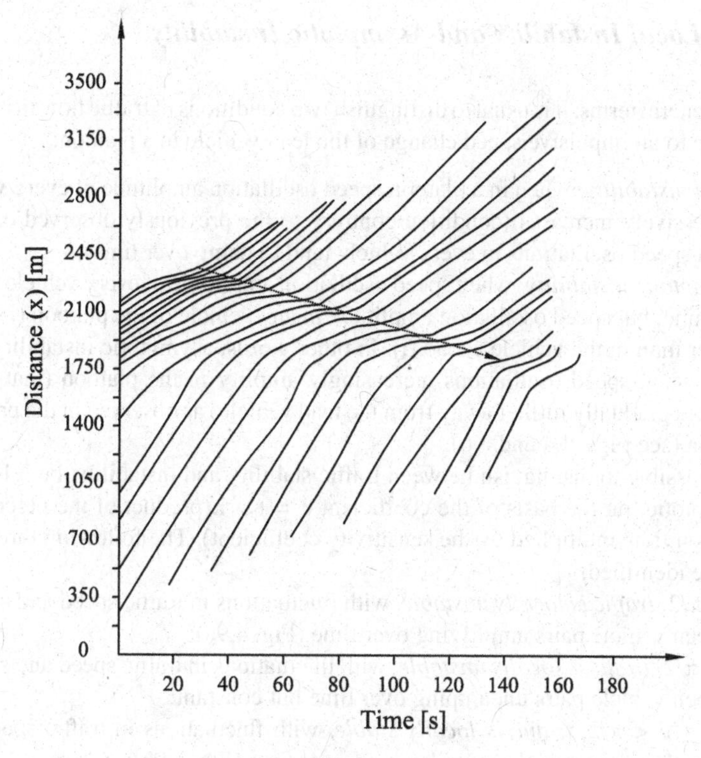

Fig. 4.7 Example of flow-breakdown (the front of the congested area tends to move backwards over time)

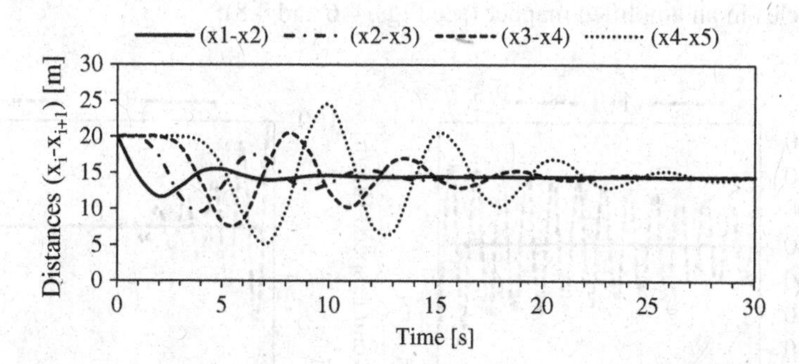

Fig. 4.8 Oscillation of distances between vehicle pairs in a platoon due to impulsive reduction of the lead vehicle speed (motorway)

In short, users can experience traffic jam conditions even in the absence of apparent reasons, that is, physical restrictions (e.g. accidents or roadworks). It is just for this reason that the described phenomenon is also defined *phantom traffic jams.*

4.3.1 Local Instability and Asymptotic Instability

In very general terms, it is usual to distinguish two conditions of traffic flow instability amenable to an impulsive speed change of the lead vehicle in a platoon:

- *local instability:* when in a platoon speed oscillation amplitude of every vehicle progressively increases (condition contrary to the previously observed cases in which speed oscillations in every vehicle tend to damp over time);
- *asymptotic instability:* when speed oscillation amplitude of every vehicle damps over time, but speed oscillation amplitude in any vehicle of the platoon (i + 1) is greater than in the vehicle ahead (i). In other words, asymptotic instability takes place when speed oscillations increasingly amplify in the platoon (that is, for vehicles gradually further away from the lead vehicle) as observed in the previous section (see Figs. 4.4 and 4.6).

It is possible to distinguish between traffic stability and instability, be it local or asymptotic, on the basis of the coefficient $c = t_{pr} \cdot \lambda$(product of the perception-reaction time multiplied by the sensitivity coefficient). The following thresholds can be identified:

- $c > \pi/2$, *traffic is locally unstable*, with fluctuations in traffic speed and spacing between vehicle pairs amplifying over time (Fig. 4.9a);
- $c = \pi/2$, *traffic is locally unstable*, with fluctuations in traffic speed and spacing between vehicle pairs undamping over time but constant;
- $1/e \leq c < \pi/2$, *traffic is locally stable,* with fluctuations in traffic speed and spacing between vehicle pairs damping over time (Fig. 4.9b);
- $c > ½$, *traffic is asymptotically unstable*, with fluctuations in traffic speed and spacing between vehicle pairs propagating backwards through a platoon of vehicles in an amplified manner (see Figs. 4.6 and 4.8);

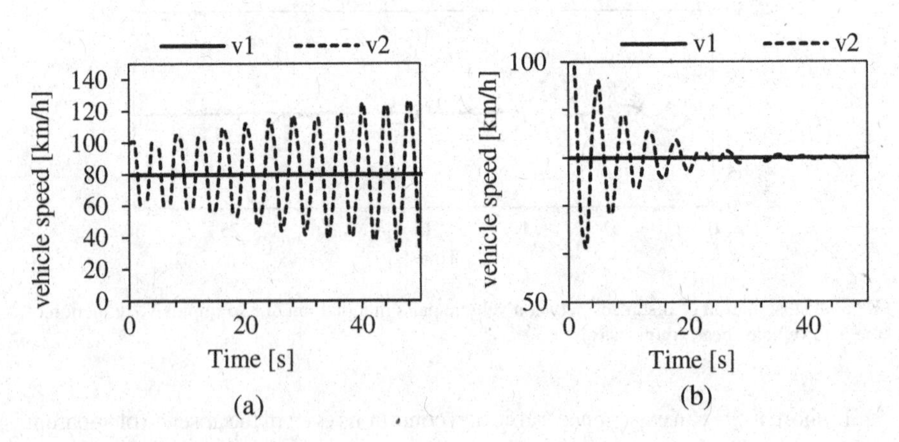

Fig. 4.9 Locally unstable traffic (left: $c > \pi/2$) and locally stable traffic (right: $1/e < c < \pi/2$)—speed of the leader vehicle (impulsive reduction of v_1 from 100 to 80 km/h) and the follower vehicle (v_2 oscillates)

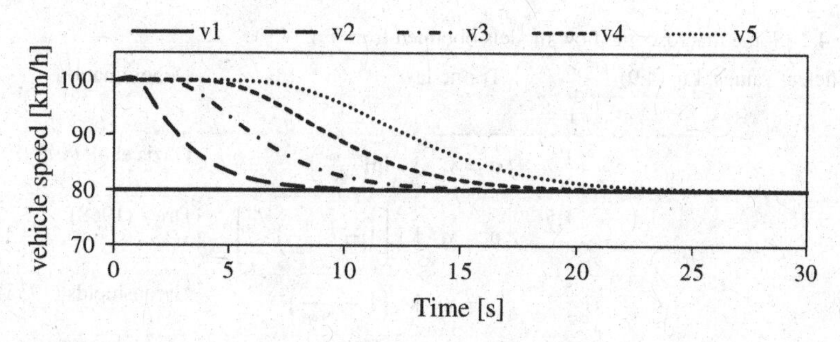

Fig. 4.10 Stable flow ($c < 1/e$) variation in the vehicle speeds in a platoon, in which the lead vehicle decelerates from 100 to 80 km/h instantaneously

- $0 < c < 1/e$, *traffic is stable*, with no fluctuations in speed and spacing between vehicles (see Fig. 4.10).

Experimental observations exclude the formation of a locally unstable flow in real cases, which thus is only a theoretical result of the car-following model. In real life, there always occurs $c < \pi/2$ (with average values closer to $c = 1/e = 0.37\ s^{-1}$) and therefore the only observable condition is asymptotic instability.

4.4 Non-linear Model

Unlike the linear model (Sect. 4.2), in the non-linear model the sensitivity λ is not constant, but has varying values as a function of the travel speed of the follower vehicle and of the spacing between vehicles (see Fig. 4.1) [4, 5]:

$$\lambda = \lambda_0 \cdot \frac{v_{i+1}^{m}(t + t_{pr})}{\left[x_i(t) - x_{i+1}(t)\right]^{l}} \tag{4.8}$$

By inserting the expression of λ in relation (4.5), it gives:

$$a_{i+1}(t + t_{pr}) = \lambda_0 \cdot v_{i+1}^{m}(t + t_{pr}) \cdot \frac{v_i(t) - v_{i+1}(t)}{\left[x_i(t) - x_{i+1}(t)\right]^{l}} \tag{4.9}$$

Coefficients λ_0 m and l allow for calibrating and particularising the model. Assuming, for instance, that $m = l = 0$ and $\lambda_0 = \lambda = \text{cost}$ gives expression (5), that is, the linear model. In the literature a plethora of other non-linear models are available (Helly [6]; Gabard et al. [7]; Gipps [8]; Bekey [9], among the others) and not mentioned here for the sake of synthesis.

Table 4.2 Some macroscopic flow models obtained from Eq. (4.9)

Coefficient values, Eq. (4.9)		Traffic law	Denomination
m	l		
0	1	$q = v_c \cdot k \cdot \ln\left(\frac{k_{jam}}{k}\right)$	Gazis et al. (1959)
0	1.5	$q = v_f \cdot k \cdot \left[1 - \left(\frac{k}{k_{jam}}\right)^{1/2}\right]$	Drew (1968)
0	2	$q = v_f \cdot k \cdot \left(1 - \frac{k}{k_{jam}}\right)$	Greenshields (1935)
1	2	$q = v_f \cdot k \cdot \exp \cdot \left(\frac{k}{k_c}\right)$	Edie (1961)
1	3	$q = v_f \cdot k \cdot e^{-\frac{1}{2}\left(\frac{k}{k_c}\right)^2}$	Drake et al. (1967)

4.5 Derivation of Macroscopic Models from the Microscopic Non-linear Model

It has been demonstrated that for the steady state the macroscopic traffic laws (see Chap. 2) can be deduced directly from the non-linear car-following model (Eq. 4.9) [4], by giving precise values to coefficients m and l as shown in Table 4.2 [10]:

4.6 Traffic Model Genealogy

In the light of the comments above, traffic models can well derive, directly or indirectly, from Greenshields' pioneering studies in 1935 (Chap. 2) which led to identify the fundamental *flow diagram* (FD). That said, they can be classified according to their detail level:

- *Macroscopic models* (see Chaps. 2 and 3), which assimilate vehicle stream to one-dimensional constant fluid on which the aggregate variables—density, speed and flow—can be determined at any road section and at any instant of time. The time evolution of these variables can be studied with the flow continuity equation (the LWR model, see Chap. 3). In short, macroscopic models describe traffic dynamic at an aggregate level;
- *Microscopic models,* which examine the motion of every vehicle in a traffic stream (speed, acceleration, driver perception-reaction time, etc.) and interactions between single vehicles. Alongside car-following models, this class encompasses *Cellular Automata—CA* models (also named *Particle Hopping* models) suggested first in 1948 [11]. Among the best-known *CA* models is *Nagel and Schreckenberg's* [12, 13];
- *Mesoscopic models,* which are characterised by an intermediate detail level with respect to micro and macro models. These models allow to analyse not the

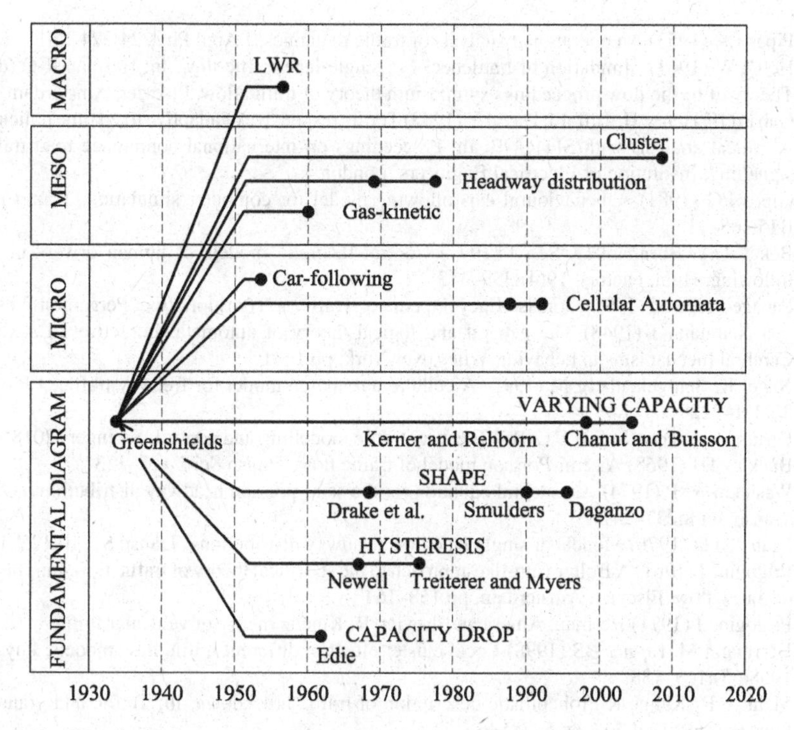

Fig. 4.11 Genealogical tree of the main traffic models and their classification

behaviour of single vehicles but rather the movement of user groups called "vehicle packages". Every vehicles package is treated as an individual entity. The packages can be modelled as punctual entities (users are assumed to be concentrated on a single point) or continuous entities (uniform user distribution between two points identifying the head and tail of the package). Among the mesoscopic models, there are three subclasses: *headway distribution models* [14–16], *gas-kinetic models* [17, 18] and *cluster models* [19, 20].

Figure 4.11 shows the genealogy of the main traffic models, subdivided into the micro-, meso- and macroscopic classes mentioned above [21].

References

1. Gerlough DL, Huber MJ (1975) Traffic flow theory; a monograph. Special report 165. TRB
2. Gipps PG (1986) A model for the structure of lane-changing decisions. Transp Res Part B Method 20(5):403–414
3. Ahmed KI, Ben-Akiva ME, Koutsopoulos HN, Mishalani RG (1996) Models of freeway lane changing and gap acceptance behavior. Pergamon, Transportation and Traffic Theory
4. Gazis DC, Herman RC, Potts RB (1959) Car-following theory of steady state flow. Oper Res 7:499–505

5. Pipes LA (1953) An operational analysis of traffic dynamics. J Appl Phys 24:274–281
6. Helly, W (1961) Simulation of bottlenecks in single-lane traffic flow. In: Herman R C (ed.) Theory of traffic flow proceedings symposium theory of traffic flow, Elsevier, Amsterdam
7. Gabard JF, Henry JJ, Tuffal J, David Y (1982) Traffic responsive or adaptive fixed time policies? A critical analysis with SITRA-B. In: Proceedings on international conference road traffic signalling, Institution of Electrical Engineers. London
8. Gipps PG (1981) A behavioural car-following model for computer simulation. Tramp Res B15:105–11
9. Bekey GA, Burnham G, Seo J (1977) Control theoretic models of human drivers in car following. Hum. Factors. 19(4):399–413
10. Papageorgiou M (1991) Concise Encyclopedia of Traffic & Transportation. Pergamon Press
11. von Neumann J (1948) The general and logical theory of automata. In: Jeffress LA (ed.) Cerebral mechanisms in behavior. Wiley, NewYork, pp 1–41
12. Nagel K, Schreckenberg M (1992) A cellular automaton model for freeway traffic. J Phys I 2:2221–2229
13. Ferrara A, Sacone S, Siri S (2018) Freeway traffic modelling and control. Springer (2018)
14. Buckley DJ (1968) A semi-Poisson model of traffic flow. Transp Sci 2:107–133
15. Wasielewski P (1974) An integral equation for the semi-poisson headway distribution model. Transp Sci 8:237–247
16. Branston D (1976) Models of single lane time headway distributions. Transp Sci 10:125–148
17. Prigogine I (1961) A Boltzmann-like approach to the statistical theory of traffic flow. In: Theory of traffic flow. Elsevier. Amsterdam, pp 158–164
18. Prigogine I (1971) Herman. American Elsevier, R. Kinetic theory of vehicular traffic
19. Herrmann M, Kerner BS (1998) Local cluster effect in different traffic flow models. Phys A 1968(55):163–188
20. Mahnke R, Kühne R Probabilistic description of traffic breakdown. In: Traffic and granular flow, Springer, vol 207, pp 527–536
21. van Wageningen-Kessels F, van Lint H, Hoogendoorn S (2015) Genealogy of traffic flow models. EURO J Transp Logist 4:445–473

Chapter 5
Fundamentals of Random and Traffic Processes

Abstract This chapter deals with the fundamentals of random and traffic processes and provides a very accurate description of probability models for arrival, speed, headway and vehicular density processes. Also the main counting, headway and speed probability distributions are shown.

Traffic flow is influenced by a wide range of factors: time variations in traffic demand; geometric (planimetric and altimetric) characteristics of infrastructures; users' behaviour (selection of routes, vehicle headways and adopted speeds); flow regulation systems (traffic lights, give-way signs etc.); interferences with pedestrian crossings etc.

Thus, in order to describe and analyse traffic flows, in a great many cases phenomena are explained with probability functions.

In other terms, very often the traffic behaviour is not perfectly foreseeable but has random nature. Several probability distributions (discrete or continuous) are applied for analysing processes concerning:

- vehicle arrival in a generic road cross section of abscissa x;
- time headway between vehicle pairs of a certain traffic flow observed in a road section x;
- vehicle density in a road segment;
- operating speed[1] (v_{85}) adopted by users in a road cross section of abscissa x.

In the following sections there will be a description of some of the most widely known probability laws used in Traffic Engineering. They are especially applied to

[1] Operating Speed is the speed at which drivers are observed operating their vehicles during free-flow conditions. The 85th percentile of the distribution of observed speeds is the most frequently used measure of the operating speed associated with a particular location or geometric feature. (AASHTO Green Book, 2001).

M. Guerrieri and R. Mauro, *A Concise Introduction to Traffic Engineering*,
Springer Tracts in Civil Engineering,
https://doi.org/10.1007/978-3-030-60723-4_5

speed processes (see Chap. 6), queuing theory (Chap. 8) and in the study of road intersections (Chap. 9).

5.1 Traffic Processes

The analysis of traffic processes is a prerequisite for describing traffic probability in an exhaustive manner.

A traffic process is a succession of measurements of a specific flow variable carried out during a time period ΔT. In other terms, a traffic process is a historical, finite-length series of measurements of a given flow variable. These measurements can concern a road cross section or segment.

The flow variables generally taken into account are the following:

- vehicle counting;
- time headway between vehicles;
- vehicle density;
- operating speed.

Figure 5.1 exemplifies the vehicle arrival counting related to measurements taken in a road cross section in a time interval ΔT. The times t_1. t_2 … t_6 represent the instants at which a vehicle passes. Each passage is marked on the time axis.

The number of the impulses in ΔT clearly represents the number of passages through the observation period ΔT. For instance, in Fig. 5.1 six vehicles have passed in the time interval ΔT.

As noted in Chap. 1, the relationship between the vehicle number m crossing the road section x in ΔT is termed *flow rate* or simply *flow* q(x):

$$q(x) = \frac{n(x)}{\Delta T} \tag{5.1}$$

If the interval ΔT is subdivided into consecutive equal time subintervals Δt_i $i = 1, 2, 3 \ldots m$, all of the same value Δt, it follows that $\Delta T = \sum_i \Delta t_i = m \cdot \Delta t$.

If the number $N(\Delta t_i) = n_i$ of vehicles passing at each Δt_i is calculated with (5.11), it yields the following sequence:

$$q_i = \frac{n_i}{\Delta t_i} \quad i = 1, 2, 3, \ldots, m \tag{5.2}$$

Equation (5.2) defines the flow rate process. The Cartesian representation of (5.2) is a piecewise-linear function obtained by attributing the i-th flow rate q_i to instant $t_i = \sum_{k=1}^{i} \Delta t_k$. $i = 1, 2, 3, \ldots, m$.

Figure 5.2 represents an example of flow rate process detected during the observation interval ΔT of around 90 min. ΔT was subdivided into sub-intervals of length

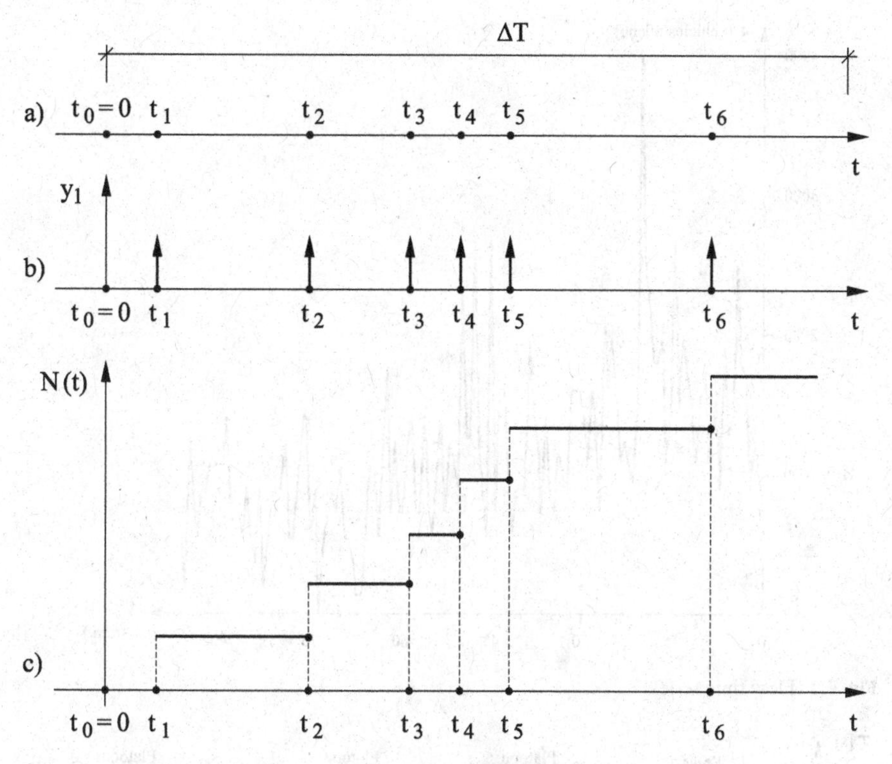

Fig. 5.1 Random flow of events and vehicle counting process on a single lane: **a** flow of events; **b** connected impulsive signal; **c** counting process (or counter process)

$\Delta t = 20$ s. The random flow rate is of crucial importance in highway engineering. It allows rigorously defining the traffic flow [1].

Figure 5.3 and Table 5.1 concern vehicle headway process.

The time headways were measured between the vehicle pairs passing through the observational cross section in succession.

The i-th headway between the i-th vehicle and (i + 1)-th of the transit sequence is shown in Table 5.1 and in Fig. 5.3 with τ_i . τ_i is attributed to the (i + 1)-th passed vehicle.

From the taken measurements a piecewise-linear function is obtained for the headways related to 60 vehicles observed in 5 min.

When recording the sequence of the instantaneous speeds v_1, v_2, ... v_n adopted by the vehicles passing through the observational cross section 'x' for a time ΔT, the speed process is obtained and represented by a straight piecewise-linear segments in Fig. 5.4.

The same Fig. 5.4 shows the process of speed levels represented by the dotted piecewise-linear segments which is given by the arithmetic mean $\bar{v}_i$ of speeds adopted up to the i-th passage and indicated next to the i-th passage, since

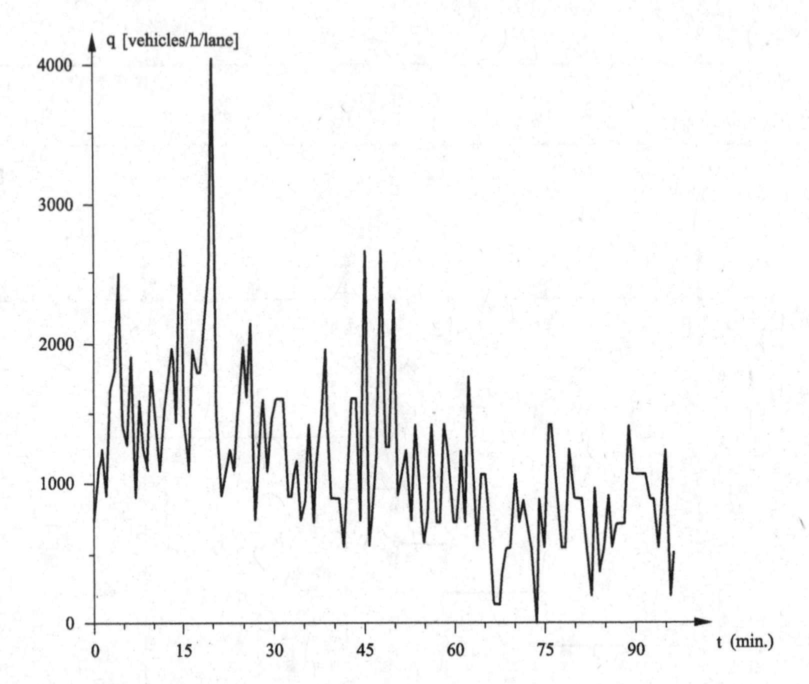

Fig. 5.2 Flow time series

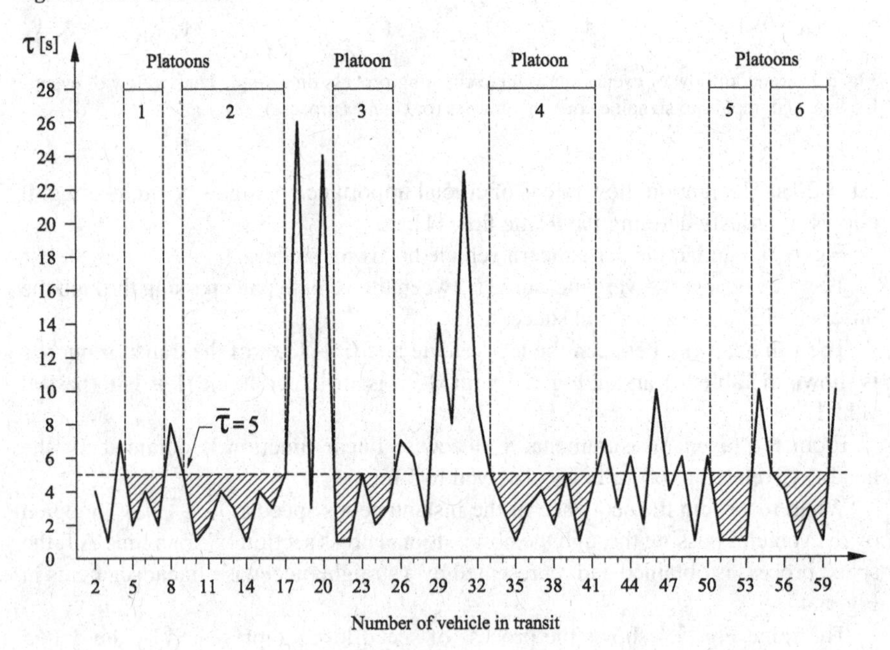

Fig. 5.3 Piecewise-linear of vehicle time headways corresponding to dataset of Table 5.1

Table 5.1 Time headway dataset (modified by source [1])

Measurement n°	τ_i (s) (between veh. i and veh. i + 1)	Measurement n°	τ_i (s) (between veh. i and veh. i + 1)	Measurement n°	τ_i (s) (between veh. i and veh. i + 1)	Measurement n°	τ_i (s) (between veh. i and veh. i + 1)
1	4 (1–2)	16	5 (16–17)	31	9 (31–32)	46	4 (46–47)
2	1 (2–3)	17	26 (17–18)	32	5 (32–33)	47	6 (47–48)
3	7 (3–4)	18	3 (18–19)	33	3 (33–34)	48	1 (48–49)
4	2 (4–5)	19	24 (19–20)	34	1 (34–35)	49	6 (49–50)
5	4 (5–6)	20	1 (20–21)	35	3 (35–36)	50	1 (50–51)
6	2 (6–7)	21	1 (21–22)	36	4 (36–37)	51	1 (51–52)
7	8 (7–8)	22	5 (22–23)	37	2 (37–38)	52	1 (52–53)
8	5 (8–9)	23	2 (23–24)	38	5 (38–39)	53	10 (53–54)
9	1 (9–10)	24	3 (24–25)	39	1 (39–40)	54	5 (54–55)
10	2 (10–11)	25	7 (25–26)	40	4 (40–41)	55	4 (55–56)
11	4 (11–12)	26	6 (26–27)	41	7 (41–42)	56	1 (56–57)
12	3 (12–13)	27	2 (27–28)	42	3 (42–43)	57	3 (57–58)
13	1 (13–14)	28	14 (28–29)	43	6 (43–44)	58	1 (58–59)
14	4 (14–15)	29	8 (29–30)	44	2 (44–45)	59	10 (59–60)
15	3 (15–16)	30	23 (30–31)	45	10 (45–46)	60	–

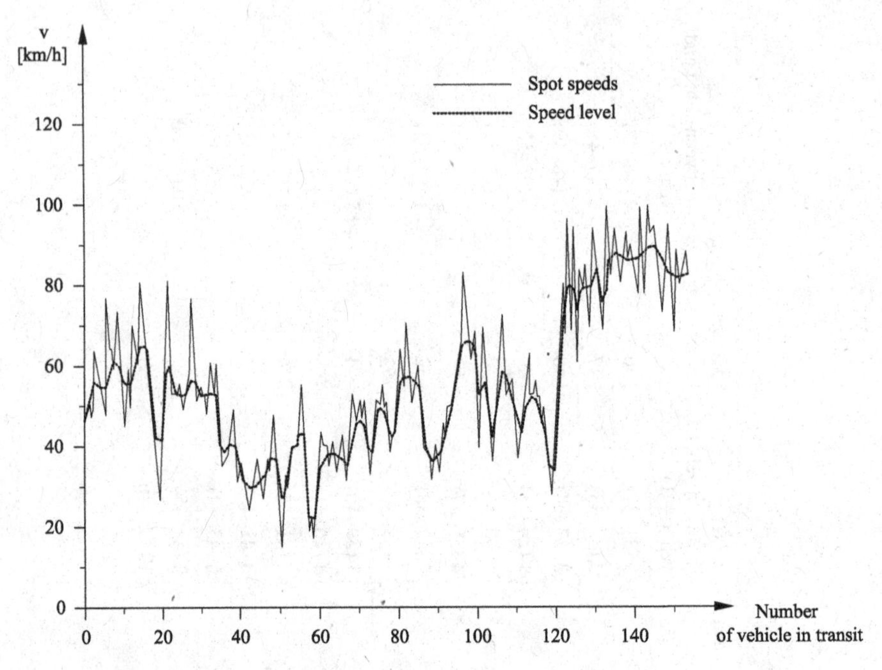

Fig. 5.4 Realization of speed process (spot speeds) and speed level

$$\bar{v}_i = \frac{\sum_{j=1}^{i} v_j}{i} \quad i = 1, 2, \ldots, n \tag{5.3}$$

Refer now to a vehicle localisation on a road segment of length L, observed at a time instant t (Fig. 5.5). Be m the number of these vehicles.

The instantaneous density at time t, defined in Chap. 1, is:

$$k = \frac{m}{L} \tag{5.4}$$

Equation (5.4) defines the vehicle density process. More precisely, this process is defined by the correspondence of densities k_1. k_2 … k_r observed in segment L at time intervals t_1. t_2 … t_r.

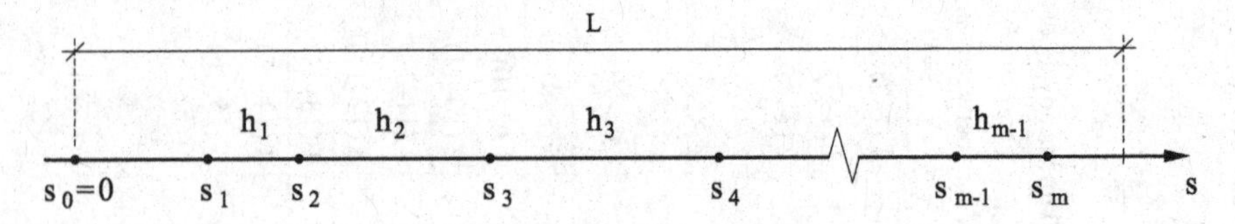

Fig. 5.5 Vehicle location and spacing along a road segment of length L

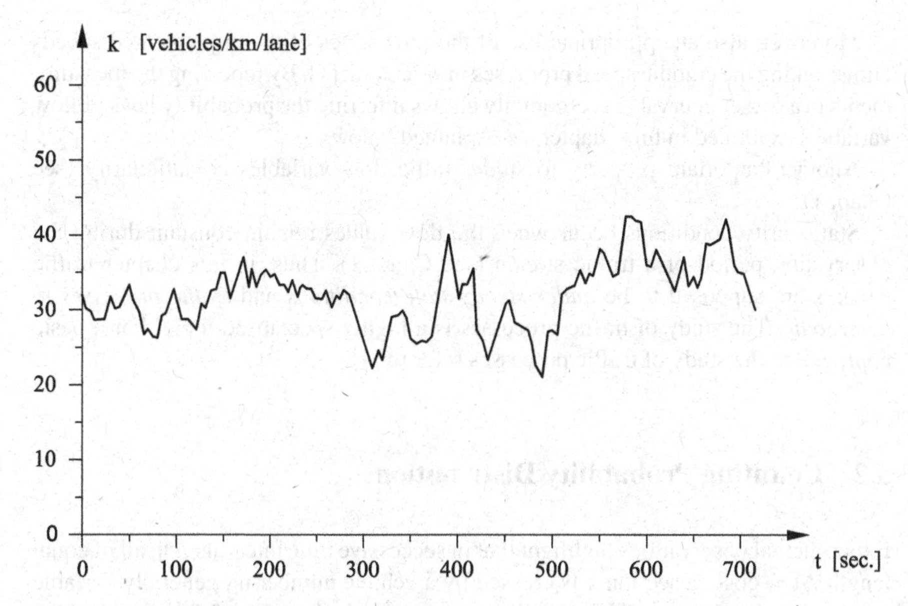

Fig. 5.6 Realisation of vehicle density processes monitored in a motorway cross section

In applications density data are indirectly inferred by measuring other traffic flow variables. The latter measurements refer to time intervals of short length (Δt equal to a few minutes) and to an elementary Δx segment straddling the cross section of abscissa x (see Fig. 1.2. Chap. 1).

In other terms, once flow q and speed v have been measured in the section x, the density k can be calculated by applying the relation (1.32) provided in Chap. 1.

Figure 5.6 shows an example of the random process referred to vehicle densities recorded in successive length intervals $\Delta t = 5$ s, on a motorway segment of given length L.

Since a traffic process is a time series of a specific traffic flow variable, it represents the realisation of a stochastic process referred to the same traffic flow variable in a finite time interval ΔT. This stochastic process is said "*random traffic process*" and constitutes the model of the homologous traffic process.

Especially in traffic engineering applications, random processes are supposed to be expression of *ergodic random processes.*

A random process is ergodic when it is stationary and its peculiar property is that a single realisation of the same process, with appropriate time interval, contains information about the entire process.

In case of ergodic process, depending on traffic conditions, successive, sufficiently numerous 'i' observations allow obtaining the specific univariate probability law for the detected traffic flow variable.

In the following sections the succession of vehicle arrivals and headways over time is supposed to be ergodic.

Moreover, also an appropriate use of the time series of the implemented speeds allows taking the ergodic speed processes into account [1]. By repeating the measurements in a preset interval ΔT, ergodicity allows inferring the probability laws of flow variables examined in this chapter, as explained below.

Another important property to study traffic flow variables is stationarity (see Chap. 1).

Stationarity conditions occur when the flow values remain constant during the observation period of a traffic stream (see Chap. 1). Thus, in this chapter traffic streams are supposed to be *under steady-state conditions* and *traffic processes to be ergodic*. The study of traffic processes is a highly specialised topic. For a basic approach to the study of traffic processes refer to [1].

5.2 Counting Probability Distributions

Experimental observations highlight that in successive time intervals Δt_i, all of equal length $\Delta t = \text{cost.}$, a section x is crossed by a vehicle number n_i, generally variable from interval to interval. If the representation of arrivals in Fig. 5.1 is repeated for each interval Δt_i ($i = 1, 2 \ldots m$), "m" determinations of the observed counting process follow. The number "n_i" of vehicles crossing the section "x" at a generic interval Δt constitutes the determination of the random variable $N(x, \Delta t)$ "number of vehicles observed on the cross section x in the time interval of length Δt"[2].

Example 1 In two successive time intervals $\Delta t_1 = \Delta t_2 = 60$ s, on the section of abscissa $x = 10$ metres there crossed 20 veh $= n_1$ in the former interval and 30 veh. $= n_2$ in the latter interval. The random variable $N(x, \Delta t)$ is thus determined as follows:

- for Δt_1: $N(10, 60) = n_1 = 20$ veh.
- for Δt_2: $N(10, 60) = n_2 = 30$ veh.

Now consider an uninterrupted flow under steady-state conditions (see Chap. 1).

Should the arrival process be ergodic, from successive "i" observations, each of length Δt, it may be possible to draw the specific univariate probability law $P[N(x, \Delta t)]$ of the traffic counts observed in the cross section of abscissa x under given traffic conditions.

If this is the case, the graphical representation of the arrival process is given by the set of "i" diagrams similar to that in Fig. 5.1, displayed the ones after the others with a common time axis.

Varying traffic conditions determine different types of probability laws $P[N(x, \Delta t)]$.

The most common probability mass functions (PMFs) used in traffic engineering are illustrated in Table 5.2 [1–3].

[2]In this case the interval ΔT in Fig. 5.1 is here set equal to Δt.

Table 5.2 Common counting distributions

Distribution	Binomial	Poisson	Negative Binomial
PMFs	$\binom{r}{n} p^n (1-p)^{r-n}$	$\frac{\mu^n}{n!} e^{-\mu}$	$\binom{n+K-1}{K-1} p^K (1-p)^n$
Mean μ	$r \cdot p$	μ	$\frac{K \cdot (1-p)}{p}$
Variance σ^2	$r \cdot p \cdot (1-p)$	μ	$\frac{K \cdot (1-p)}{p^2}$
$\frac{\mu}{\sigma^2}$	$(1-p)^{-1} > 1$	1	$p < 1$
Parameter estimation	$p = (\bar{n} - s^2)/\bar{n}$ $r = \bar{n}^2/(\bar{n} - s^2)$	$\mu = \bar{n}$	$p = \bar{n}/s^2$ $K = \bar{n} \cdot p/(1-p)$
Traffic condition	Congested traffic	Light traffic	Cyclic variation in flow

In Table 5.2 $\bar{n}$ represents the mean of the number of vehicles that pass the cross section x calculated in observation time intervals.

In traffic engineering there is widespread use of generalized Poisson distribution defined by two real and positive parameters K and λ, with the following expression when K is integer:

$$P(N = n) = \sum_{i=1}^{K} \frac{e^{-\lambda} \cdot (\lambda)^{n \cdot K + i - 1}}{(n \cdot K + i - 1)!} \quad \text{with } n = 0, 1, 2 \ldots \tag{5.5}$$

It may be assumed as a first approximation that $K = \bar{n}/s^2$ being $\bar{n}$ the sample mean and s^2 the sample variance of the variable $N(x, \Delta t)$ (that is of the sample of determinations of vehicle passages along the road section x during the time intervals Δt).

Between the two parameters K and λ and the mean $\bar{n}$ there is the following relation:

$$\lambda = K \cdot \bar{n} + \frac{1}{2} \cdot (K - 1) \tag{5.6}$$

K and λ can be evaluated by means of the nomograph in Fig. 5.7, provided by Haight [2], in function of the mean and variance values.

For vehicular flows up to 1500 veh/h the K values in Table 5.3 can be used as reference. Afterwards, the Eq. (5.6) allows the calculation of λ (data displayed in Table 5.3 as well) [1].

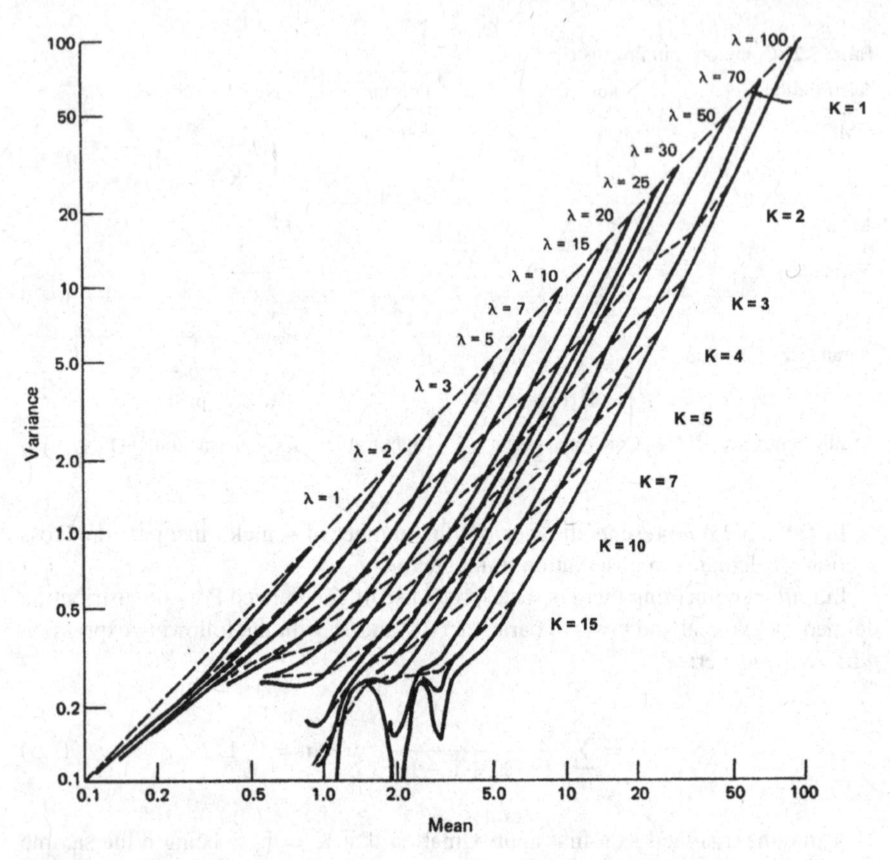

Fig. 5.7 Nomograph to calculate parameters K and λ of the generalized Poisson distribution function [2]

Table 5.3 K and **λ** values as a function of q to be used in the generalized Poisson distribution

Flow q [veh/h]	K value	λ value
0–500	1	$\lambda = \bar{n}$
501–1000	2	$\lambda = 2 \cdot \bar{n} + 1/2$
1001–1500	3	$\lambda = 3 \cdot \bar{n} + 1$

5.2.1 *General Criterion for Selecting the Appropriate Counting Probability Distribution*

In order to identify the appropriate probability distribution for a specific traffic analysis, it needs calculating the ratio between the sample mean $\bar{n}$ and the sample variance s^2 of the variable $N(x, \Delta t)$. In other words, $\bar{n}$ and s^2 are calculated on the sample of determinations of vehicle passages through the cross section x during the time intervals Δt.

Should the sample mean $\bar{n}$ and the sample variance s^2 be thought as similar to each other $\bar{n} \approx s^2$, it may be assumed that the observed statistical distribution conforms to a Poisson law. For it the theoretical mean μ and the variance σ^2 are actually equal to the same value, which also describes the entire model. Poisson's distribution is generally valid for relatively modest flows, that is, for $q(x) = 400$–500 veh/h.

On the other hand, should the sample mean be higher than the sample variance, it can be inferred that the variability in measures is lower than that expected for purely Poissonian arrivals of equal mean. In this case the concordance of experimental data may be validated with either the positive binomial or the generalised Poisson model described above, which have a higher mean than the variance. In this case, μ and σ^2 also define completely the two probability distributions.

The circumstance that $\bar{n} > s^2$ has been frequently observed in traffic streams far away from the free flow condition (Chap. 2), when flows are generally high. The generalized Poisson distribution can be used as a counting law also for high flow values provided that the flow uniformity (Chap. 1) is maintained during the observation period of arrivals.

Finally, if the mean is lower than, or equal to the variance, there are more dispersed counts than those deriving from purely Poissonian arrivals. This being the case, the general option is a negative binomial distribution, otherwise known as Pascal's. Again, mean and variance completely identify the model.

Traffic counts resulting in $\bar{n} < s^2$ and in conformity to Pascal's law are mostly observed under flow conditions downstream of traffic lights, even if some researchers have used such a distribution as an arrival model on multi-lane roads.

Table 5.4 sums up the criterion for selecting the models just described on the basis of measured values of mean $\bar{n}$ and variance s^2 [1]:

Whatever the PMF identified with the criterion provided above, it always needs to carry out a specific statistical conformity test (e.g. *chi-square test*) in order to evaluate the likelihood of data fitting ("comparison" between estimated data with the chosen PMF and the data inferred from the traffic surveys).

Considering Eq. (1.2) of Chap. 1, for the probability distributions in Table 5.2 the parameter mean μ of vehicle arrivals at the cross section x during the time interval Δt is given by:

$$\mu = q(x) \cdot \Delta t = \bar{n} \tag{5.7}$$

Table 5.4 Criterion for selecting the counting model

Ratio between variance s^2 and mean	Distribution suggested
$s^2/\bar{n} > 1$	Negative binomial
$s^2/\bar{n} \approx 1$	Poisson
$s^2/\bar{n} < 1$	Binomial or Generalised Poisson

By way of an example, the probability that in a section x, with a flow q, n vehicles cross the section during a time interval Δt with a Poissonian PMF (see Table 5.2), is as follows:

$$P[N = n] = \frac{(q \cdot \Delta t)^n \cdot e^{-q \cdot \Delta t}}{n!} \tag{5.8}$$

With the same PMF, the probability that lower than n vehicles pass, is given by the following expression:

$$P[N \leq n] = \sum_{n_i=0}^{n_i=n} \frac{(q \cdot \Delta t)^{n_i} \cdot e^{-q \cdot \Delta t}}{n_i!} \tag{5.9}$$

Example 2 In a motorway with three-lane carriageways 1800 vehicles were measured in an hour on a carriageway cross section. Assuming the flow as constant over time and a Poissonian PMF, determine: (a) the probability that 3 vehicles pass in an interval $\Delta t = 10$ s; (b) the probability that less than 4 vehicles pass in the same time interval; (c) the probability that no vehicle passes in a time interval $\Delta t = 5$ s.

For the problem (a) it results:

$$q = 1800 \text{ veh/h} = 1800/3600 = 0.5 \text{ veh/s}$$

$$P[N = 3] = \frac{(0.5 \cdot 10)^3 \cdot e^{-0.5 \cdot 10}}{3!} = 0.14$$

For the problem (b) it results:

$$P[N \leq 4] = \sum_{n_i=0}^{n_i=3} \frac{(0.5 \cdot 10)^{n_i} \cdot e^{-0.5 \cdot 10}}{n_i!} = 0.27$$

For the problem (c) it results:

$$P[N = 0] = \frac{(0.5 \cdot 5)^0 \cdot e^{-0.5 \cdot 5}}{0!} = 0.08$$

Thus, in 8% of cases in 5-s intervals there is no vehicle passage.

Example 3 Table 5.5 shows a set of determinations of the number n of vehicles pass through a cross section x of a suburban highway lane, recorded in 64 intervals, each long $\Delta t = 15$ s (column I). The frequencies f_i observed for every value n_i of n are those in column II.

Table 5.5 Empirical and probability distributions of traffic

I	II	III		IV		V	
n ≡ vehicles/15 s	Observed frequencies f_i	Poisson		Binomial		Generalized Poisson	
		P(n)	Theoretical frequencies	P(n)	Theoretical frequencies	P(n)	Theoretical frequencies
$n_1 = 0$	0	0.00057	0.04	0.00004	0.00	0.00000	0.00
$n_2 = 1$	0	0.00426	0.27	0.00060	0.04	0.00014	0.01
$n_3 = 2$	0	0.01592	1.02	0.00393	0.25	0.00191	0.12
$n_4 = 3$	3	0.03962	2.54	0.01604	1.03	0.01190	0.76
$n_5 = 4$	0	0.07398	4.74	0.04563	2.92	0.04290	2.75
$n_6 = 5$	8	0.11052	7.07	0.09588	6.14	0.10056	6.44
$n_7 = 6$	10	0.13757	8.80	0.15389	9.85	0.16524	10.58
$n_8 = 7$	11	0.14678	9.39	0.19247	12.32	0.20071	12.85
$n_9 = 8$	10	0.13703	8.77	0.18956	12.13	0.18741	11.99
$n_{10} = 9$	11	0.11372	7.28	0.14751	9.44	0.13864	8.87
$n_{11} = 10$	9	0.08493	5.44	0.09040	5.79	0.08324	5.33
$n_{12} = 11$	1	0.05767	3.69	0.04317	2.76	0.04136	2.65
$n_{13} > 12$	1	0.07743	4.95	0.02088	1.33	0.02599	1.65
	64	1.00000	64.00	1.00000	64.00	1.00000	64.00

For these measured distributions, it results:

$$\sum_{i=1}^{13} n_i \cdot f_i = 478 \text{ veh}$$

$$\sum_{i=1}^{13} n_i^2 \cdot f_i = 3822 \text{ veh}$$

and, thus, from the numerousness of the sample $N = \Sigma f = 64$ are given the values of the sample mean $\bar{n}$ and the sample variance s^2 of the number of vehicles passing the cross section under study[3] in $t = 15$ s:

$$\bar{n} = qt = \frac{478}{64} = 7.469 \text{ veh /15 s}$$

$$s^2 = \frac{3822 - (\overline{478}^2/64)}{63} = 3.999 \text{ (veh /15 s)}^2$$

[3]For calculating s^2 the relation between the moments of a frequency distribution was used, on whose basis $s^2 = (n_2^* - \bar{n}^2/N)/(N - 1)$ where $n_2^* = \sum n_i^2 \cdot f_i$ is the second moment of the frequency distribution.

With regard to the second member of the mean $\bar{n}$, it is worth noting that the quantity q is the average flow in the time unit (here $q = 7.469/15 = 0.498$ veh/s) and Δt is the length of the intervals in which the passing vehicles are counted (in this case $\Delta t = 15$ s).

The relation $\bar{n} = q \cdot \Delta t$ is in agreement with 5.7) since, whatever PMF is chosen to interpret data, $\bar{n}$ represents the estimation of the counting model mean. Resulting the ratio between sample mean and variance as:

$$\frac{\bar{n}}{s^2} = \frac{7.469}{3.999} = 1.868$$

i.e. higher than the unity, it is possible to select the binomial probability distribution law and particularise it using the observed data.

This probability law (PMF) depends on the two parameters p and r; r is provided by

$$r = \frac{\bar{n}^2}{\bar{n} - s^2} = \frac{(7.469)^2}{7.469 - 3.999} = 16.077$$

approximating to $r = 16$, since for the binomial PMF r must be a positive integer number.

For p it yields

$$p = \frac{\bar{n} - s^2}{\bar{n}} = \frac{\bar{n}}{r} = \frac{7.469}{16} = 0.467$$

If we put $q = 1 - p = 1 - 0.467 = 0.533$ starting from the general expression:

$$P(N = n) = \binom{r}{n} p^n (1 - p)^{r-n} = \frac{r!}{n!(r - n)!} p^n q^{r-n} \quad n = 0, 1, 2, \ldots$$

the wanted law assumes the form:

$$P(N = n) = \binom{16}{n} (0.467)^n (0.533)^{16-n} = \frac{16!}{n!(16 - n)!} (0.467)^n (0.533)^{16-n}$$

with mean

$$\mu = r \cdot p = 16 \cdot 0.467 = 7.472 \text{ veh/15 s}$$

and variance

$$\sigma^2 = r \cdot p \cdot q = 16 \cdot 0.467 \cdot 0.533 = 3.999 \text{ (veh/15 s)}^2$$

In alternative to the binomial law, should the circumstance be $\bar{n}/s^2 > 1$, the generalized Poisson distribution can be adopted as a theoretical model to be then subjected to a conformity test.

In order to identify such distribution, K and λ parameters are to be estimated from the available data, in that they entirely complete it and are linked to each other by relation (5.6).

Since by using the values of $\bar{n} = 7.469$ veh/15 s and $s^2 = 3.999$ (veh/15 s)^2 from the nomograph in Fig. 5.7 it follows $K = 2$, from (5.6) it yields:

$$\lambda = 2 \cdot \bar{n} + 0.5 = 2 \cdot 7.469 + 0.5 = 15.438$$

From the general expression of the generalised Poisson law

$$P(N = n) = \sum_{i=1}^{K} \frac{e^{-\lambda} \cdot (\lambda)^{n \cdot K + i - 1}}{(n \cdot K + i - 1)!} \quad n = 0, 1, 2, \ldots$$

whose mean and variance cannot be generally expressed in closed form, the following PMF is particularized on the data under consideration

$$P(N = n) = \sum_{i=1}^{2} \frac{e^{-15.438} \cdot (15.438)^{2 \cdot n + i - 1}}{(2 \cdot n + i - 1)!}$$

Finally, although the relation $\bar{n}/s^2 > 1$ does not recommend the adoption of a Poissonian PMF as a model, it is applied to the data under consideration for its key role in several Highway Engineering issues. The only parameter which fully defines this law is the mean μ, thus yielding

$$P(N = n) = \frac{\mu^n}{n!} e^{-\mu} \quad n = 0, 1, 2, \ldots$$

Moreover, for the variance it follows $\sigma^2 = \mu$.

In case of the counts in Table 5.5 the P(n) particularizes into

$$P(N = n) = \frac{(7.469)^n}{n!} e^{-7.469}$$

with mean $\mu = 7.469$ veh/15 s and variance $\sigma^2 = 7.469$ (veh/15 s)^2. The value of the latter entirely disagrees with the estimation provided by the observations equal to $s^2 = 3.999$ (veh/15 s)^2.

However, through Poisson counting distribution, probabilities related to a surveyed arbitrary count m_i result to be those in column III of Table 5.5, which shows the theoretical frequencies $64 \cdot P(n)$ together with probability distributions obtained by the models in question.

By comparing the theoretical with the experimental frequencies f_i in column II, it is easily observable that the Poisson law unsatisfactorily approximates the statistical series (n, f); on the other hand, the binomial and the generalised Poisson models provide more accurate frequencies. The latter generalized Poisson model can thus be an alternative to the most common binomial count model.

A quantitative evaluation of the gap between theoretical and empirical series is made only by a conformity test (e.g. chi-square or Kolmogorov-Smirnov goodness-of-fit tests).

Finally, the study aims at comparing the probability distributions in Table 5.2 in case of equal mean. Thus, the probabilities for the random variable 'arrival of N vehicles in a 20-s interval' when N assumes the following values $N = 1, 2, 3, \ldots, 11$ or more vehicles, supposing a traffic flow $q = 720$ veh/h $= 0.2$ veh/s .

By using the symbols and relations of Table 5.2, it follows:

- Poisson probability distribution

yielding

$$\mu = 0.2 \cdot 20 = 4 \text{ veh/20 s}; \ \sigma^2 = 4 \text{ (veh/20 s)}^2$$

and therefore

$$P(n) = \frac{\mu^n \cdot e^{-\mu}}{n!} = \frac{4^n \cdot e^{-4}}{n!}$$

- binomial probability distribution

supposing the unit interval (20 s) be subdivided into twenty equal parts of 1 s, it follows:

$$\mu = r \cdot p = 4 \text{ veh/20 s}$$

from which

$$p = 0.2 \ 1 - p = 0.8 \ \sigma^2 = r \cdot p \cdot (1 - p) = 3.2 \text{ (veh/20 s)}^2$$

and thus

$$P(n) = \frac{20!}{(20 - n)!n!} \cdot 0.2^n \cdot 0.8^{20-n}$$

- negative binomial probability distribution

given the μ value, assuming $K = 2$, it follows

$$\mu = K \cdot (1 - p)/p \text{ or } 4 = 2 \cdot (1 - p)/p$$

from which

$$p = 0.3334;\ 1 - p = 0.6666;\ \sigma^2 = \frac{2 \cdot 0.6666}{(0.3334)^2} = 12(\text{veh}/20\,\text{s})^2$$

These values yield

$$P(n) = \binom{n+K-1}{K-1} \cdot p^K \cdot (1-p)^K = \frac{(n+1)!}{n!} \cdot (0.3334)^2 \cdot (0.6666)^n$$

- generalized Poisson distribution

Given the value of $q = 720$ veh/h; $K = 2$ is chosen (Table 5.3), thus resulting in

$$\bar{n} = \mu = 4\ \text{veh}/20\,\text{s}$$

and, therefore, always from Table 5.3

$$\lambda = K \cdot \bar{n} + \frac{1}{2}(K-1) = 8.5$$

Through this determination, in general, for (5.5) it yields

$$P(n) = \sum_{i=1}^{2} \frac{e^{-\lambda} \cdot \lambda^{K \cdot n + i - 1}}{(K \cdot n + i - 1)!} = e^{-\lambda} \cdot \left[\frac{\lambda^{2n}}{(2n)!} + \frac{\lambda^{2n+1}}{(2n+1)!}\right]$$

$$= e^{-8.5} \cdot \left[\frac{8.5^{2n}}{(2n)!} + \frac{8.5^{2n+1}}{(2n+1)!}\right]$$

The values of probabilities P(n) are summed up in Table 5.6 and graphed in Fig. 5.8.

Table 5.6 Probability values according to different distributions with equal mean

P(N = n)	Poisson	Binomial	Negative binomial	Generalized poisson
P(N = 0)	0.0183	0.0115	0.1112	0.0019
P(N = 1)	0.0732	0.0576	0.1483	0.0277
P(N = 2)	0.1464	0.1369	0.1482	0.1175
P(N = 3)	0.1952	0.2054	0.1318	0.2320
P(N = 4)	0.1952	0.2182	0.1098	0.2628
P(N = 5)	0.1562	0.1746	0.0878	0.1924
P(N = 6)	0.1041	0.1091	0.0683	0.0982
P(N = 7)	0.0595	0.0545	0.0520	0.0369
P(N = 8)	0.0297	0.0222	0.0390	0.0106
P(N = 9)	0.0132	0.0074	0.0289	0.0024
P(N = 10)	0.0053	0.0020	0.0212	0.0004
P(N = 11 or more)	0.0037	0.0006	0.0535	0.0172

5.3 Probability Distribution for Time Headways

As well as the counting process, also the traffic process of time headways between vehicles (see Sect. 5.2) is of particular interest to applications.

Hereafter the random variable "vehicle time headway between vehicle pairs" is denoted with τ; τ is a continuous-type random variable.

Below the traffic random process is assumed to be ergodic and a renewal process. A renewal process takes place when under steady-state conditions the random variables $\tau_1, \tau_2, \ldots \tau_n$ turn out to be independent from one another.

The univariate probability density functions (PDFs) most used in highway engineering are those shown in Table 5.7 [1, 4].

In the same Table $\bar{\tau}$ and s^2 respectively denote the sample mean and the sample variance of the determinations of the headways τ_i between pairs of successive vehicles passing the cross section x and belonging to a flow under steady-state conditions ($q(x) = q$).

It is proved that [1]:

- if vehicle passages are distributed according to a Poissonian PMF (i.e., for $q(x) =$ 400–500 veh/h), the headway τ is an exponential random variable. The Poisson probability model is to be selected (and later verified through a conformity test) if the sample mean and the sample mean squared deviation have a similar value, or: $\bar{\tau} \cong \sqrt{s^2}$;
- for very high values of flow (small average headways), τ is a log-normal random variable;
- if two or more flows with the same or opposite travel direction are considered and the flow value is medium/high, an Erlang distribution may be assumed for τ.

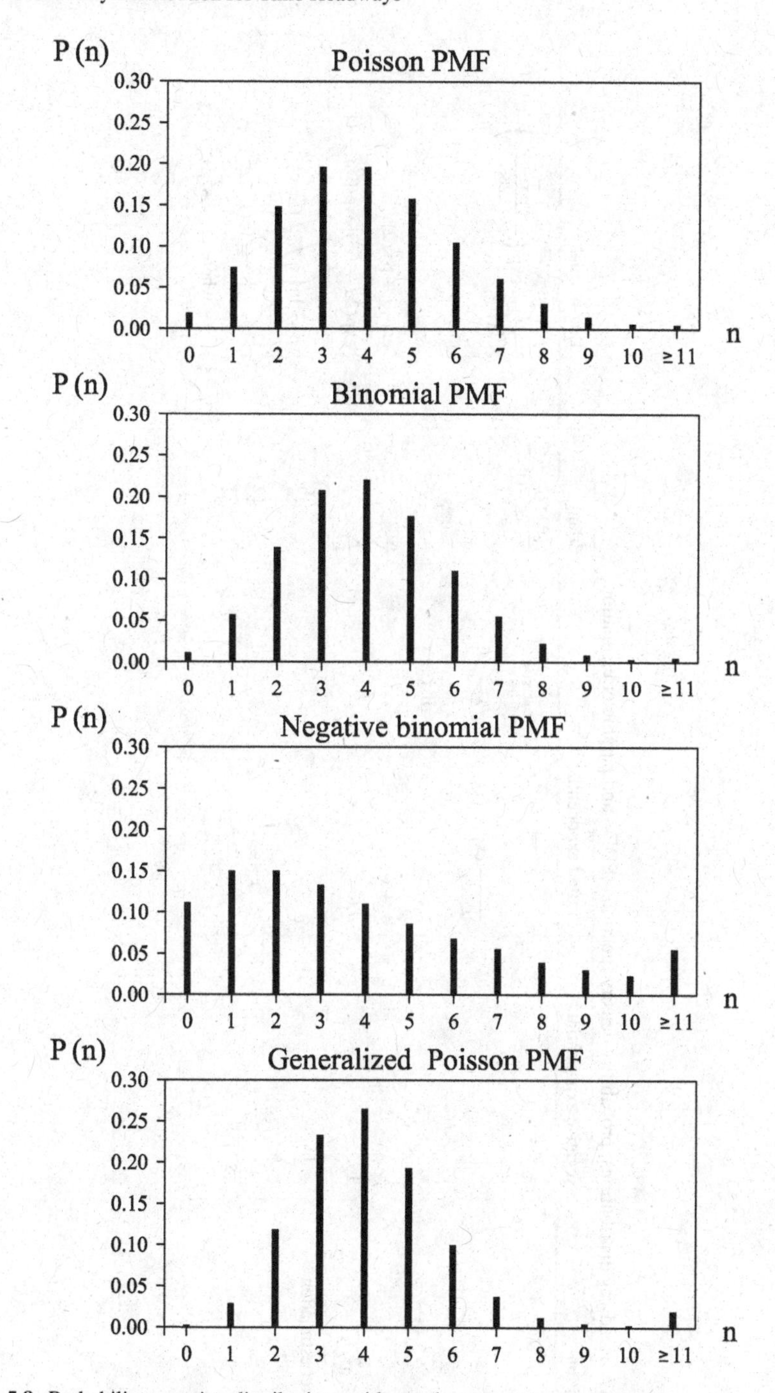

Fig. 5.8 Probability counting distributions with equal mean

Table 5.7 Headway distributions: probability density functions (PDFs) and parameter estimation

Distribution	Negative exponential	Shifted exponential	Erlang	Log-normal
PDF $f_\tau(\tau)$	$\lambda e^{-\lambda\tau}$	0 if $\tau < c$ $\frac{1}{\frac{1}{\lambda} - c}\exp[-(\tau - c)/(1/\lambda - c)]$ if $\tau \geq c$	$\frac{\lambda e^{-\lambda\tau}(\lambda\tau)^{K-1}}{(K-1)!}$	$\frac{1}{\tau\beta\sqrt{2\pi}}\exp\left(-\frac{(\ln\tau-\alpha)^2}{2\beta^2}\right)$
Mean μ	$\frac{1}{\lambda}$	$\frac{1}{\lambda} + c$	$\frac{K}{\lambda}$	$\exp(\alpha + \beta^2/2)$
Variance σ^2	$\frac{1}{\lambda^2}$	$\frac{1}{\lambda^2}$	$\frac{K}{\lambda^2}$	$\exp(2\alpha + \beta^2)[\exp(\beta^2) - 1]$
Parameter estimation	$\frac{1}{\lambda} = \bar{\tau}$ $\frac{1}{\lambda^2} = s^2$	$c = \bar{\tau} - \sqrt{s^2}$ $\lambda^2 = \frac{1}{s^2}$	$\lambda = \frac{\bar{\tau}}{s^2}$ $K = \frac{\bar{\tau}^2}{s^2}$	$\alpha = \ln\left(\bar{\tau} \Big/ \sqrt{1 + c_\tau^2}\right)$ $\beta^2 = \ln(1 + c_\tau^2)$ where $c_\tau^2 = s^2/\bar{\tau}^2$

Table 5.8 Distribution functions of different RVs

Probability distribution	Cumulative distribution function
Negative exponential	$F_\tau(\tau) = 1 - e^{-q \cdot \tau}$
Shifted exponential	$F_\tau(\tau) = 1 - e^{-[(\tau - c)/(1/q - c)]}$
Erlang	$F_\tau(\tau) = 1 - e^{Kq\tau} \sum_{z=0}^{K-1} \frac{(Kq\tau)^z}{z!}$

Bearing in mind Eq. 1.6 in Chap. 1, under steady-state flow conditions it follows that the mean μ numerically corresponds to the reciprocal of the flow:

$$\mu = \bar{\tau} = \frac{1}{q} \tag{5.10}$$

Thus, for instance, the exponential PDF particularises as follows, being $\lambda = q$ (see Table 5.7):

$$f_\tau(\tau) = q \cdot e^{-q \cdot \tau} \tag{5.11}$$

The probability P that the vehicle headway may have a value between two threshold levels ($\tau_1 \leq \tau \leq \tau_2$), is calculated by the cumulative distribution function (CDF) $F_\tau(\tau)$. It yields:

$$P(t_1 \leq t \leq t_2) = F_t(t_2) - F_t(t_1) \tag{5.12}$$

Depending on the random variable (R.V.) the cumulative distribution functions are expressed as in Table 5.8 [1].

Figures 5.9a, b display the behaviours of the Erlang PDF when K varies in two cases connected to the parameter λ. Figure 5.9a assumes $\lambda = \cos t$, consequently the distribution mean increases linearly with K. Figure 5.9b displays the K-influence in case of constant mean, that is $K/\lambda = \cos t$. The latter case is of interest to model headways.

$$\mu = K/\lambda = 4\,s$$

Example 4 n values of vehicle headways were measured in a road cross section, ($\tau_1 = 3.03$ s; $\tau_2 = 3.83$ s; … $\tau_n = 2.97$ s). Thus, the sample mean and sample variance were calculated as follows: $\bar{\tau} = 3.52$ s and $s^2 = 5.04$ seconds^2. Assess the flow and the probability that the headway may be: a) lower than 3.20 s.; b) between 3.20 and 3.80 s; c) higher than 3.80 s.

- The flow is (see Eq. 5.10): $q = 1/\bar{\tau} = 1/3.52 = 0.28$ veh/s $= 1008$ veh/h;

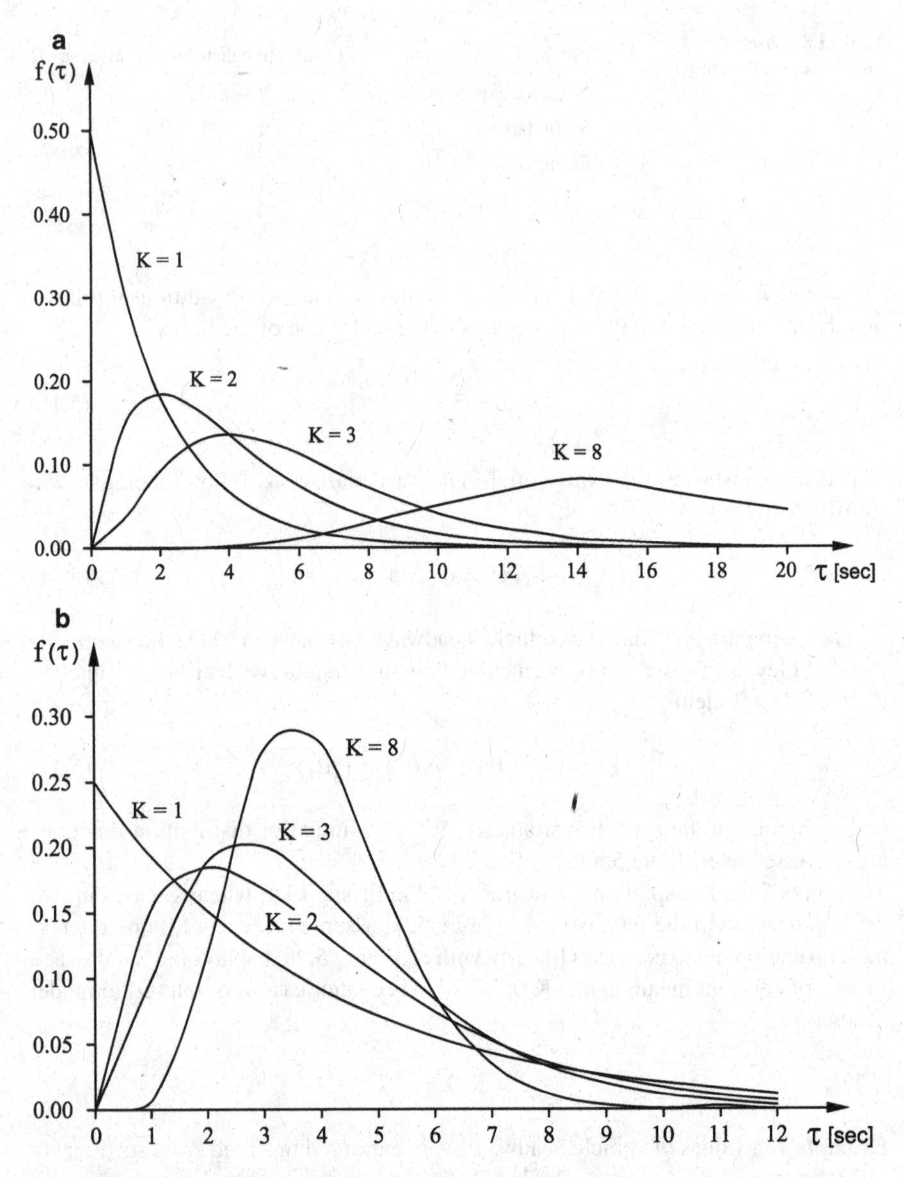

Fig. 5.9 **a** Erlang probability density function with $\lambda = 0.5$ and some K values. **b** Erlang probability density function for some K values and with a constant mean

- Not being $\bar{\tau} \cong \sqrt{s^2}$, there is no chance for the PDF to be exponential but rather a shifted exponential, except for being subjected to a subsequent conformity test. For the shifted exponential it follows (see Table 5.7):

$$l = 1/\sqrt{s^2} = 1/\sqrt{5.04} = 0.44 \text{ veh/s}$$

$$c = \bar{\tau} - \sqrt{s^2} = 3.52 - \sqrt{5.04} = 1.28 \text{ s}$$

The distribution function in Table 5.8 particularises as follows:

$$F_\tau(\tau) = 1 - e^{-[(\tau-c)/(1/q-c)]} = 1 - e^{-[(\tau-1.28)/(3.52-1.28)]} = 1 - e^{-0.44\cdot(\tau-1.28)}$$

The probabilities that the vehicle headways are lower than 3.20 and 3.80 s are as follows:

$$P(\tau \leq 3.20) = F_\tau(3.20) = 1 - e^{-0.44\cdot(3.20-1.28)} = 0.57$$

$$P(\tau \leq 3.80) = F_\tau(3.80) = 1 - e^{-0.44\cdot(3.80-1.28)} = 0.67$$

The probability that the headway ranges between 3.20 s and 3.80 s is:

$$P(3.20 \leq \tau \leq 3.80) = F_\tau(3.80) - F_\tau(3.20) = 0.67 - 0.57 = 0.10$$

The probability that the headway is higher than 3.80 s is as follows:

$$P(\tau > \tau^*) = 1 - F_\tau(\tau^*); \text{ thus:}$$

$$P(\tau > 3.80) = 1 - F_\tau(3.80) = 1 - 0.67 = 0.33$$

5.4 Speed Processes

Let v be the instantaneous speed performed by the single vehicle crossing a certain section x. The random variable V, of which v is a realization, has a probability distribution that in many cases can be considered of the normal type.

The PDF f(v) is entirely defined after estimating the mean μ and variance σ^2 parameters respectively, as follows:

$$f(v) = \frac{1}{\sqrt{2\pi}\cdot\sigma}\cdot\exp\left[-\frac{(v-\mu)^2}{2\sigma^2}\right] \tag{5.13}$$

The conditions of the road transport system affect the shape of the curve f(v), in that they contribute to form the values μ e σ^2.

For instance, in Fig. 5.10 there are two instantaneous speed probability density functions, (a) and (b), which are in line with the corresponding statistical distributions

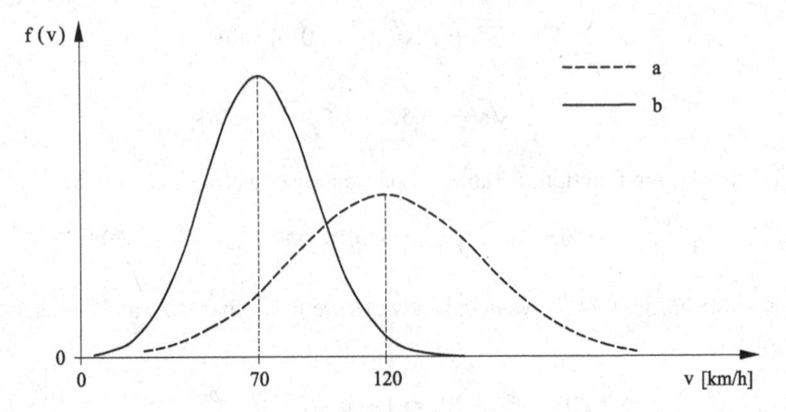

Fig. 5.10 Examples of instantaneous speed probability density functions

observed in the same road section. The probability density function (a) concerns low flow, consisting only of cars and good environmental conditions. The probability density function (b) regards high mixed flow of cars and commercial vehicles, in the rain.

In the heavy-traffic conditions (case b), the instantaneous speeds are distributed around a central value ($\mu = 70$ km/h) and a dispersion (σ) which are both lower than the model concerning case a).

These circumstances are thought to occur in the following conditions (ideal configuration of road and traffic):

- low flow values (free flow);
- motion of single unrelated vehicle towards the road alignment;
- day and good weather conditions.

The comparison between the shapes of the desired speed distributions with those of speeds performed in non-ideal conditions can provide concise information on the effects which globally such situations produce on drivers by conditioning their behaviours.

In addition to normal distribution, other PDFs [1] have been proposed. They are represented by the log-normal and by compound probability densities (dichotomic processes).

The log-normal PDF was described in Sect. 5.3 as specific probability law for vehicle headways. Indeed, it is equally useful for interpreting the instantaneous speeds when the flow conditions reduce most of the performed v values in a small number of classes. On the other hand, the remaining v values are distributed along an endless right queue.

The compound probability functions were introduced, like for headways (see Sect. 5.3), to describe traffic streams composed of vehicle sets which can be characterised as either free or conditioned or subject to varying speed limits (e.g. on light or heavy vehicles).

Consequently, different PDF combinations may exist depending on vehicle composition and flow regime of the traffic streams examined. For instance, a dichotomic probability law for v can derive from a linear combination $f(v) = b \cdot g(v) + (1 - b) \cdot h(v)$ of a normal PDF (es. g(v)) and a log-normal PDF for either subsystem of heavy vehicles (in percentage b and (1 − b) of the whole traffic).

Should statistical distributions be bimodal, the general pragmatic approach is adopting a compound law to model performed speeds. If this is the case, it is convenient to distinguish and identify the two PDFs composing the dichotomic law on research.

Finally, it goes without saying that the process of performed speeds has long been the object of systematic surveys which are extremely interesting in motorways.

Let ..., V_{t-1}, V_t, V_{t+1}, ... be the random variables representing the speed of vehicles of a traffic stream passing through a certain location x in the instants ... $t - 1, t, t + 1$, ... The sequence of V_t is the random process of the traffic stream speeds.

Let ..., v_{t-1}, v_t, v_{t+1}, ... be a realization of this random process (see Fig. 5.4) and $\bar{v}_t$ the average value of the realizations of the variables V_t up to the instant t. The difference between v_t and $\bar{v}_t$ is also a realization of a random variable α_t denoted a_t; therefore, by experimentation the following relationships have been verified:

$$v_t = \bar{v}_t + a_t \tag{5.14}$$

$$\bar{v}_t - \bar{v}_{t-1} = \lambda \cdot a_t \tag{5.15}$$

Actually, a_t are the determinations of independent and identically distributed random variables with zero mean and variance σ^2.

The coefficient λ has values between 0 and 1. Actually, λ measures the influence that the speed fluctuations recorded at a certain instant have on average on the speed of the next passing vehicle at the observed location x.

From the Eqs. (5.14) and (5.15) we get:

$$v_t = \bar{v}_{t-1} + \lambda \cdot a_{t-1} + a_t \tag{5.16}$$

If we replace $\bar{v}_{t-1}$ in Eq. (5.16) with the expression obtained from Eq. (5.14) for $t - 1$, we have:

$$v_t = v_{t-1} + a_t - a_{t-1} \cdot (1 - \lambda) \tag{5.17}$$

The Eq. (5.17) represents a first order integrated moving average process (ARIMA).

When the traffic flow is low and the vehicles movement does not occur in platoons, vehicles travel along the lane at very large distances on average. In this case, λ is practically equal to zero. From Eq. (5.15) we obtain that the level of $\bar{v}_t$ is constant over time. Equation (5.14) shows that the speeds measured at the observed location x for the progressive t, are identically distributed Gaussian random variables, with constant mean and variance equal to that of a_t. In this case the sequence of the v_t is the realization of a renewal process.

When the traffic flow is not low, it turns out that $\lambda > 0$. In this case the sequence of the v_t is the realization of a process with equation:

$$v_t = v_{t-1} + a_t - (1 - \lambda) \cdot a_{t-1} \quad (5.18)$$

The process that occurs according to Eq. (5.18) depends on λ and σ^2. As said, λ and σ^2 are functions of the spacing between vehicles.

The processes described above are of great importance in estimating the reliability of traffic streams. These aspects are summed up in the following Chap. 6.

References

1. Mauro R (2015) Traffic and random processes. Springer, Berlin
2. Gerlough DL, Huber MJ (1975) Traffic flow theory: a monograph. Special Report 165. TRB
3. Leutzbach W (1988) Introduction to the theory of traffic flow. Springer, Berlin
4. Teodorović D, Janić M (2016) Transportation engineering. Theory: practice and modeling. Elsevier, Amsterdam

Chapter 6
Traffic Management and Control Systems

Abstract This chapter presents an advanced method for estimating flow reliability on highways and describes the current systems of highway traffic management and control. In this chapter a rigorous capacity definition is offered. The basic characteristics of the Automated Highway System are also introduced. Finally, the HSM method to estimate the annual crash frequency expected on highways and the COPERT method to calculate the polluting emissions are both illustrated.

6.1 Preliminary Considerations

Flow instability is a major problem in highway engineering. As a matter of fact, instability can turn into flow-breakdown which is (see Chap. 4) a random phenomenon, in that it depends on such stochastic factors as:

- vehicle speed variations (due to lane changes or highway on-ramps);
- driver perception/reaction time (t_{pr});
- driver sensitivity coefficient λ.

According to Eq. (4.8), the coefficient λ increases when the distance between vehicle pairs decreases. In accordance with relation (1.12) the mean space headway $\overline{h}_s$ is equal to the reciprocal of the vehicle density k. Thus, the probability that an instability condition occurs, increases with the increasing frequency in speed variations of the stream and density k.

The techniques for flow-breakdown risk control are based on either direct measurements of the traffic flow reliability or comparison of the calculated density with a threshold value k*.

k* is the reference value against which the flow breakdown risk should be considered as inacceptable [1].

In order to reduce the frequency of flow instability occurrences, the first step may be to either limit lane change numbers of vehicles driving along the same carriageway

M. Guerrieri and R. Mauro, *A Concise Introduction to Traffic Engineering*,
Springer Tracts in Civil Engineering,
https://doi.org/10.1007/978-3-030-60723-4_6

or reduce the highway on-ramp input flows. This in fact tends to keep the density k below the set threshold k*.

Among the density control strategies, there are:

- highway *on-ramp metering*;
- temporary permit for using the emergency or hard shoulder lane (thus also favouring increase in capacity).

The most known and used speed control strategies are:

- truck overtaking prohibition (i.e. prohibition to travel on the overtaking lane for heavy vehicles);
- *variable speed limits* (VLS).

In the former case, the advantage is to avoid sudden speed reductions on the overtaking lane due to the temporary presence of heavy vehicles.

In the latter case, on the other hand, imposing the same maximum speed limit in all lanes (e.g. 80 km/h) and below the ordinary one (e.g. 130 km/h) reduces the frequency in lane changes. This is consequent to the fact that users, due to speed limits, cannot modify their speed.

Moreover, users see the drivers on the other lanes travelling with the same speed as theirs. Consequently, they have no advantage of shifting from a lane to another.

6.2 Flow Reliability on Highways

Examine a cross section of abscissa x of a road with divided carriageways (e.g. motorways and highways) in an observation time ΔT. ΔT is subdivided into a number "m" of sub-intervals, all of the same value Δt_i (Δt_1, Δt_2, ... Δt_m), for instance 5′ or 15′. With specific reference to the overtaking lane, in each interval Δt_i can be measured and calculated:

- the vehicle number crossing the section x, $n_i(x)$, and flow $q_i(x)$ (see Eq. (1.2));
- the time mean speed $\overline{v}_{t,i}(x)$ (see Eq. (1.7)).

The counting ($n_1(x)$, $n_2(x)$, ..., $n_m(x)$) and speed ($\overline{v}_{t,1}(x)$; $\overline{v}_{t,2}(x), \ldots \overline{v}_{t,m}(x)$) sequences are random processes. Especially the latter is the realisation of an autoregressive integrated moving average (ARIMA) of the first order process [2–4] characterized by two parameters:

- the deviation between the instantaneous speed of every vehicle crossing the cross section of abscissa x in the interval Δt_i and the average speed value $\overline{v}_{t,i}(x)$ calculated in this interval;
- a coefficient between 0 and 1 measuring the influence of the deviation between the mean speed calculated for the generic interval Δt_i (that is, $\overline{v}_{t,i}(x)$) and the speed calculated for the successive interval Δt_{i+1} (or, $\overline{v}_{t,i+1}(x)$).

The deviation is characterised by a variance σ^2. As shown by Ferrari, [2–4], there exists the following empirical relation obtained from experimentation on highways:

$$\sigma^2 = M \cdot \ln(k_i) + N \tag{6.1}$$

where

- M and N are model constants ($M < 0$ and $N > 0$);
- k_i is the vehicle density in the overtaking lane calculated in the interval Δt_i; it yields $k_i = q_i(x)/\overline{v}_{t,i}(x)$.

Observations have proved that the varied regression lines $\sigma^2 = M \cdot \ln(k_i) + N$ pass all through the same coordinate point: $\sigma^2 = 1.5\ m^2/s^2$ and $k = 30$ veh/km [3, 5].

Ferrari's research has also shown that the flow instability condition is a random event linked to the process of the mean speeds reached by drivers; the instability starts from the overtaking lane to extend to the other lanes. It was observed in Chap. 2 that when density increases, the space mean speed decreases; when the density assumes values near k_{jam}, the space mean speed tends to zero. Before reaching the highest density k_{jam}, the flow is already unstable, in that vehicles are subject to several stops and restarts.

The probability that the flow instability does not occur in a given time period T is termed *flow reliability* Φ with regard to that period.

In other words, the reliability of the traffic stream, which is stable in the observation period T, is the probability that the time mean speed $\overline{v}_{t,i}(x)$ decreases up to zero in the time interval after T. Instead, in the observed period T no remarkable speed reductions take place. *Reliability* Φ is calculated with the following expression:

$$\Phi = 1 - 19.80\left(\frac{q}{10,000}\right)^{8.82} T^{1.933} M^2 \tag{6.2}$$

where

- q (veh/h) denotes the flow on the overtaking lane;
- T (minutes) is the chosen observation period;
- M (m^2 km/s^2) is the model constant.

It clearly follows $0 \leq \Phi \leq 1$. When the reliability values are near 0, there is a high probability that in the observation period T the stream is along the unstable segment of the speed-flow diagram (Fig. 2.4, Chap. 2) and that there are significant speed reductions. On the other hand, if $0 \leq \Phi \leq 0.85$, there is a good reliability of the vehicle stream. For further theoretical and application studies, the interested reader may consult [2, 3, 5–7].

With probability Φ, the flow that can cross a section of the overtaking lane in the period T with no instability phenomena is [2]:

$$q = 10{,}000 \cdot \left[\frac{1-\Phi}{19.80 \cdot T^{1.933} \cdot M^2}\right]^{\frac{1}{8.82}} \tag{6.3}$$

There follows a further definition of lane capacity which, according to the model just described, corresponds to the flow calculated with Eq. (6.3) for a reliability value near 1 (generally $\Phi = 0.90$) in a conventional period $T = 15'$. Thus, the carriageway capacity c_{carr} is equal to the total flow passing a carriageway cross section when the previously defined flow (see Eq. (6.3)) is reached on the overtaking lane. Consequently, if the carriageway is composed of n-lanes (generally $n = 2$ or 3), by denoting the flow of the ith-lane with q_i, (besides the flow on the overtaking lane), there follows:

$$c_{carr} = 10{,}000 \cdot \left[\frac{1-0.90}{19.80 \cdot 15^{1.933} \cdot M^2}\right]^{\frac{1}{8.82}} + \sum_{1=1}^{n-1} q_i \tag{6.4}$$

6.2.1 *Case Study: Assessment of the Reliability Laws from Traffic Surveys*

In a segment of a highway overtaking lane the instantaneous vehicle speeds shown in Table 6.1 were observed in a time period $\Delta T = 5$ min. Sub-intervals $\Delta t_i = 20$ s were considered in order to determine the speed process. Assess the reliability laws as the flow varies for periods of 15′, 20′ and 30′.

In order to solve the problem, the first step is to compute the mean value of the instantaneous speeds of single vehicles for every 20-s interval. The calculation values are given in the last column of Table 6.1.

Then, the deviations between the recorded instantaneous speeds and mean speed may be calculated for each interval; the values are shown in Table 6.2.

By means of the values in Table 6.1 and the calculation values in Table 6.2, the following data can be obtained:

- vehicle number crossing the segment: $n(x) = 93$ veh;
- flow $q(x) = n(x)/\Delta T = 93/5 = 18.6$ veh/min $= 1116$ veh/h; mean speed in the whole observation period $\overline{v}_t(x) = 22.0\,\text{m/s}$;
- average density $k = 1116/(22.0 \cdot 3.6) = 14.\ 1$ veh/km;
- variance of speeds $\sigma^2 = 7.6\ \text{m}^2/\text{s}^2$.

Since the line $\sigma^2 = M \cdot \ln(k_i) + N$ also crosses both the point just identified ($\sigma^2 = 7.6\ \text{m}^2/\text{s}^2$; $k = 14.1$ veh/km) and, as previously explained, the point ($\sigma^2 = 1.5\ \text{m}^2/\text{s}^2$; $k = 30$ veh/km), it follows that $M = (7.6–1.5)/(\ln 30–\ln 14.1) = 8.08\ \text{m}^2\ \text{km/s}^2$. Once known the M value, through the relation (6.2) the sought-after reliability laws can be obtained (Fig. 6.1).

Table 6.1 Instantaneous speeds observed in 20-s intervals

N°	Interval	$n_i(x)$	$v_i(x)$ (m/s)							$\overline{v}_{t,i}(x)$(m/s)
			1	2	3	4	5	6	7	
1	14:00:00–14:00:20	6	16	18	23	28	22	24		21.8
2	14:00:20–14:00:40	7	19	22	23	18	20	25	23	21.4
3	14:00:40–14:01:00	6	24	27	26	20	22	25		24.0
4	14:01:00–14:01:20	6	28	25	22	22	21	22		23.3
5	14:01:20–14:01:40	6	25	22	21	21	25	26		23.3
6	14:01:40–14:02:00	6	16	16	24	23	24	20		20.5
7	14:02:00–14:02:20	7	19	24	21	19	20	22	17	20.3
8	14:02:20–14:02:40	6	20	23	20	25	25	25		23.0
9	14:02:40–14:03:00	5	22	18	22	22	19			20.6
10	14:03:00–14:03:20	6	21	26	22	21	22	25		22.8
11	14:03:20–14:03:40	6	26	17	22	22	25	20		22.0
12	14:03:40–14:04:00	6	22	20	18	23	16	23		20.3
13	14:04:00–14:04:20	7	22	25	26	23	20	18	23	22.4
14	14:04:20–14:04:40	7	23	26	16	17	21	26	24	21.9
15	14:04:40–14:05:00	6	24	18	20	26	23	25		22.7

Table 6.2 Deviation values between instantaneous and mean speeds in every interval

N°	Interval	Deviations: $v_i(x) - \overline{v}_{t,i}(x)$ (m/s)						
		1	2	3	4	5	6	7
1	14:00:00–14:00:20	−5.8	−3.8	1.2	6.2	0.2	2.2	–
2	14:00:20–14:00:40	−2.4	0.6	1.6	−3.4	−1.4	3.6	1.2
3	14:00:40–14:01:00	0.0	3.0	2.0	−4.0	−2.0	1.0	–
4	14:01:00–14:01:20	4.7	1.7	−1.3	−1.3	−2.3	−1.3	–
5	14:01:20–14:01:40	1.7	−1.3	−2.3	−2.3	1.7	2.7	–
6	14:01:40–14:02:00	−4.5	−4.5	3.5	2.5	3.5	−0.5	–
7	14:02:00–14:02:20	−1.3	3.7	0.7	−1.3	−0.3	1.7	−4.8
8	14:02:20–14:02:40	−3.0	0.0	−3.0	2.0	2.0	2.0	–
9	14:02:40–14:03:00	1.4	−2.6	1.4	1.4	−1.6	–	–
10	14:03:00–14:03:20	−1.8	3.2	−0.8	−1.8	−0.8	2.2	–
11	14:03:20–14:03:40	4.0	−5.0	0.0	0.0	3.0	−2.0	–
12	14:03:40–14:04:00	1.7	−0.3	−2.3	2.7	−4.3	2.7	–
13	14:04:00–14:04:20	−0.4	2.6	3.6	0.6	−2.4	−4.4	1.2
14	14:04:20–14:04:40	1.1	4.1	−5.9	−4.9	−0.9	4.1	2.2
15	14:04:40–14:05:00	1.3	−4.7	−2.7	3.3	0.3	2.3	–

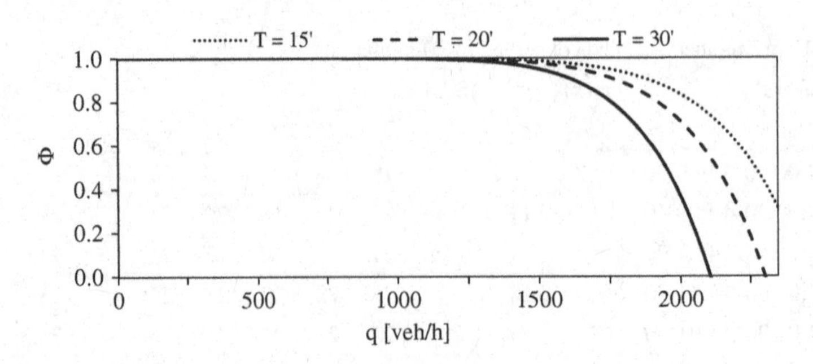

Fig. 6.1 Relations $\Phi = \Phi(q, T)$

6.3 The Ramp-Metering

The *ramp-metering* is a traffic-light control system at ramps (Fig. 6.2) designed to manage and limit the density on specific segments of a highway infrastructure. Its main objectives are:

- to keep the density below a given value, always lesser than the critical value (k_c), so as to let the flow occur in the stable portion of the fundamental diagram (see Fig. 2.4). This can be obtained by cyclically stopping the flow on the ramps with consequent number reduction in vehicles entering the highway segments;
- to avoid flow instability phenomena due to sudden vehicle speed variations on the highway, following on-ramp entries of other vehicles;
- to guarantee an adequate level of service LOS (the latter being a function of the density k, see Table 2.3), also with regard to the current traffic demand;
- to increase the flow speed and reduce the total travelling time of all drivers;
- to reduce accident risk.

Adopting such a system, however, means to accept the possibility of queue formation on ramps (which therefore need a proper geometry for vehicle accumulation),

Fig. 6.2 Examples of highways with a ramp-metering system

with a consequent increase in travelling time on the road network external to the highway.

Access control can be differently imposed [6]:

- on the basis of *techniques founded on traffic flow series*, which can be of two types:
  - *traffic-responsive control* in function of the flow conditions measured in real time by loop detectors;
  - *pre-timed control*, that is, determined in function of statistics concerning the variations in the highway flow (during daily hours, weekdays etc.);
- on the basis of the *regulation logic*, in this case the control can be imposed on single ramps (i.e. thought as isolated systems) or as a coordinated control on more ramps. Ramps can be controlled with different logics:
  (a) capacity/demand control;
  (b) occupancy control;
  (c) feed-back control termed ALINEA in isolated ramps and METALINE in more coordinated ramps;
- in function of the access regulation technique:
  (a) *for a single vehicle* if the traffic light cycle is planned to allow for the single-vehicle entry by cycle (green time + yellow time $\approx$ 3 s);
  (b) *for vehicle platoons*, if the traffic light cycle allows for the entry of two or three vehicles (green time + yellow time $\approx$ 4.7 s and 6.4 s, respectively);
  (c) *for critical intervals*: the traffic lights allow for the go-ahead when, on the highway outside lane (right lane), the traffic control system measures gaps between successive vehicle pairs higher than the critical one required for an on-ramp entry in full safety conditions.

6.3.1 System Analysis for an Isolated On-Ramp

Consider an isolated on-ramp of the type illustrated in Fig. 6.3.

Be:

- q_i the flow upstream the ramp;
- $c = q_{max}$ the carriageway capacity;

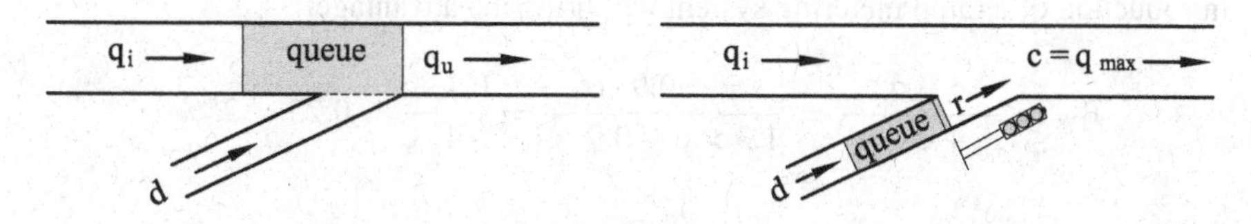

Fig. 6.3 Highway on-ramp without and with an access control system

- d the on-ramp traffic demand;
- q_u the flow downstream the ramp.

If $q_i + d > q_{max} = c$, the carriageway capacity downstream the ramp is below the sum of the flow from the upstream segment, with respect to the ramp, and the traffic demand of the ramp itself. It results in a queue formation on the ramp and jam on the highway segment upstream the ramp (with queues).

Downstream the ramp, a flow q_u below the capacity $q_u < c$ can be detected (experimentally it is observed q_u as 5–10% below the capacity) [6, 8, 9].

On the other hand, where the vehicle entry on-ramp is monitored and controlled, by allowing the flow $r = q_{max} - q_u$ to enter the highway segment downstream the ramp, the flow will be equal to capacity and no queues will develop on the highway but only on the ramp.

In order to evaluate the advantage from such a control system in terms of users' total waiting time, a deterministic queue model is applied (see Chap. 8).

By denoting the queuing vehicle number and the total waiting time with no ramp-metering in a generic time interval, with N_s and W_s respectively, the homologue traffic parameters with ramp-metering with N_c and W_c and the advantage with B_w, it yields:

$$N_s = (q_i + d - q_u) \cdot T \tag{6.5}$$

$$W_s = \int_0^T (q_i + d - q_u) \cdot t \cdot dt = (q_i + d - q_u) \cdot \frac{T^2}{2} \tag{6.6}$$

$$N_c = (q_i + d - c) \cdot T \tag{6.7}$$

$$W_c = \int_0^T (q_i + d - c) \cdot t \cdot dt = (q_i + d - c) \cdot \frac{T^2}{2} \tag{6.8}$$

$$B_W = \frac{W_s - W_c}{W_s} = \frac{c - q_u}{q_i + d - q_u} \tag{6.9}$$

Example 1 A highway has a carriageway capacity $c = 3200$ pcu/h for each direction. The total traffic demand in the segment downstream the on-ramp is $(q_i + d) = 1.3 \cdot c$. Moreover, the flow $q_u = 0.90 \cdot c$ was measured after the jam. In this case the introduction of a ramp metering system will bring the advantage:

$$B_W = \frac{c - q_u}{q_i + d - q_u} = \frac{c - 0.9 \cdot c}{1.3 \times c - 0.9 \cdot c} = \frac{0.1 \cdot c}{0.4 \cdot c} = 0.25 = 25\%$$

6.3.2 System Analysis for On-Ramps and Off-Ramps

Examine a highway segment with an off-ramp followed by an on-ramp.
Be:

- q_i the flow upstream the off-ramp;
- d the on-ramp traffic demand;
- $c = q_{max}$ the carriageway capacity;

Assume by hypothesis that the flow downstream the queue be $q_{max} = c$.

Figures 6.4 and 6.5 show two highway layouts without and with ramp-metering system, respectively. Such a system is clearly useful: vehicles off the first ramp (off-ramp) do not have the delays which instead occur in absence of ramp-metering as a consequence of the queue caused by the entry flow from the second ramp (on-ramp).

Figure 6.4 illustrates the no ramp-metering layout, which brings the following conditions:

- the flow off the first ramp q_{ex} is equal to the rate γ of the flow q_1;
- the flow off the section x_3 is $q_3 = q_{max}$. Moreover, the flow totally entering the section downstream the on-ramp is $q_2 + d$. Therefore, for the continuity law: $q_2 = q_{max} - d$;
- in the segment between the sections x_1 and x_2, it results $q_1 = q_{ex} + q_2$. The off-flow is a rate γ of q_1, or $q_{ex} = \gamma \cdot q_1$.

In the light of what specified above:

$$q_1 = q_{ex} + q_2 = \gamma \cdot q_1 + q_2 = \gamma \cdot q_1 + q_{max} - d \tag{6.10}$$

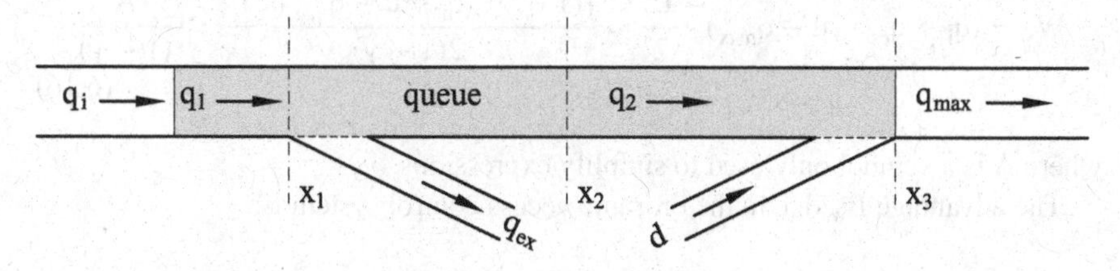

Fig. 6.4 Highway without the ramp-metering system

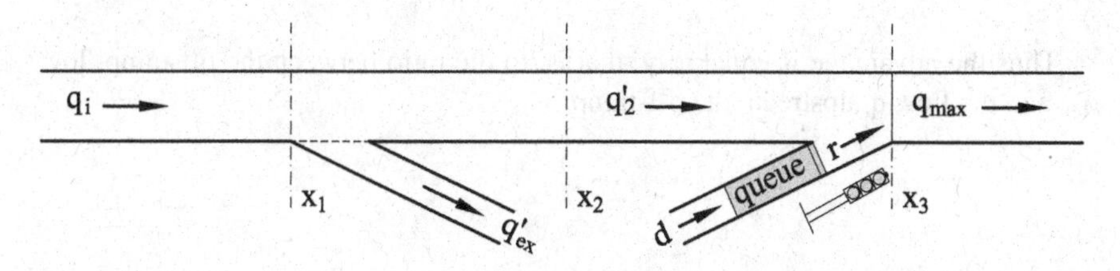

Fig. 6.5 Highway with the ramp-metering system

By applying the continuity law to the segment between the sections x_1 and x_3, the expression of q_{ex} can be obtained:

$$q_{ex} = \gamma \cdot q_1 = \gamma \cdot (q_{max} - d + q_{ex}) \tag{6.11}$$

thus:

$$q_{ex} = \frac{\gamma}{(1-\gamma)} \cdot (q_{max} - d) \tag{6.12}$$

If the ramp-metering system is active (Fig. 6.5), the whole demand d cannot enter the highway but rather a lesser flow r (r < d) so as to determine the capacity downstream the on-ramp, or:

$$r = q_{max} - q_2' \tag{6.13}$$

where

$$q_2' = (1-\gamma) \cdot q_i \tag{6.14}$$

The total waiting time in presence and in absence of the ramp-metering system, W_c and W_s respectively, can be inferred – as explained in the previous paragraph—through the deterministic queue model (see Chap. 8):

$$W_c = [(1-\gamma) \cdot q_i + d - q_{max}] \cdot \frac{T^2}{2} = A \tag{6.15}$$

$$W_s = (q_i + q_{ex} + d - q_{max}) \cdot \frac{T^2}{2} = \frac{[(1-\gamma) \cdot q_i + d - q_{max}] \cdot T^2}{2(1-\gamma)} = \frac{A}{(1-\gamma)} \tag{6.16}$$

where A is a symbol only used to simplify expressions.

The advantage B_w due to the on-ramp access control system is:

$$B_W = \frac{W_s - W_c}{W_s} = \frac{\frac{A}{(1-\gamma)} - A}{\frac{A}{(1-\gamma)}} = \gamma \tag{6.17}$$

Thus the advantage is equal to γ, that is, to the ratio between the off-ramp flow q_{ex} and the flow q_i upstream the off-ramp.

6.4 Hard Shoulder Running System

The temporary use of the hard-shoulder (or the emergency) lane, as an auxiliary traffic lane, allows to increase the carriageway capacity during traffic peaks.

This traffic management system is also termed "*dynamic lane*" or "*HSR*", (*hard-shoulder running*). It is a system used in a great many western countries in order to increase the capacity in existing highways.

Dynamic lanes are fairly widespread in Germany (since 1996, together with the *variable speed limit* of 100 km/h), in Holland (since 2003), in the United Kingdom (since 2006, together with the *variable speed limit* of 50 mph).

In Italy this system is in use on the A14 (for 23.6 km) and on "the Mestre ring road" (part of the A4 motorway).

Much more numerous are, then, the installations in the USA: in San Diego (California) since 2005 on the highway I-805/SR 52 (the system sets off when the flow speed decreases below 30 mph); in Delaware; in Florida since 2005 on Highways SR 826 and SR 836; in Georgia on Highway GA 400 since 2005 (the system sets off when the flow speed lowers below 35 mph); in Maryland on Highway US 29 (with the allowed maximum 55 mph speed); and also in New Jersey, Virginia and Washington.

Experiences on Dutch highways have showed that the hard shoulder running system leads to a 7–22% increase in carriageway capacity [10]. In Germany the following were recorded: capacity increase up to 25%, travelling time reduction up to 20% [11] and air and noise pollution reduction.

Especially in Europe (more specifically, in Germany, Holland, England), HSR are combined with variable speed limits [10] (see Sect. 6.5).

6.4.1 Capacity Estimation of Highways with HSR System

In order to estimate the benefits brought about by the activation of an HSR system in a divided highway, in terms of carriageway capacity increase, the first key step is to introduce a behavioural driver model.

A two-lane divided highway is considered here.

In a two-lane divided highway with HSR system, the carriageway capacity can be estimated by analogy with what occurs in highways with three lanes on each carriageway.

More precisely, a driver opts for the HSR lane if there is a low probability that he will reach a slower vehicle, or a high probability that, once reached, he will accept its speed, mainly without changing lanes. The central and left-hand lanes of a three-lane carriageway are basically utilised in the same way as the middle and outside (overtaking) lanes in a two-lane carriageway, respectively.

This behavioral scheme of a three-lane carriageway is consistent with the results of traffic surveys made on highway segments with two or three lanes passed by traffic flow vehicles of a comparable layout.

For instance, the highest capacities measured in a three-lane segment of the Highway A1 (before the carriageway enlargement that led to the current configuration) between Modena and Bologna were, respectively, 1902 veh/h; 1504 veh/h; 967 veh/h, starting from left to right. Instead, the capacities recorded in a two-lane carriageway of the Highway A14 between Bologna and Imola were 1889 veh/h and 1465 veh/h, again from left to right [12].

Therefore, the capacities of the central and overtaking lanes of a highway with three lanes on each carriageway can be in practice considered as coinciding, respectively, with the capacities of the right and overtaking lanes of a two-lane divided highway.

Consequently, in two-lane divided highways with HSR system, the free-flow speed $(v_f)_{right}$ and critical density $(k_c)_{right}$ of the right-hand lane (that is the hard shoulder running lane), can be estimated from the assumption that the free-flow speed and critical density of the central lane ($(v_f)_{central}$ and $(k_c)_{central}$) are the mean values of those of the right lane and overtaking lane ($(v_f)_{over}$ and $(k_c)_{over}$).

This means, with respect to $(v_f)_{right}$ and $(k_c)_{right}$, to solve these two equations

$$\frac{(v_f)_{over} + (v_f)_{right}}{2} = (v_f)_{central} \tag{6.18}$$

$$\frac{(k_c)_{over} + (k_c)_{right}}{2} = (k_c)_{central} \tag{6.19}$$

Once known $(v_f)_{right}$ and $(k_c)_{right}$, the critical speed $(v_c)_{right}$ is obtained from the chosen traffic stream model (e.g. Drake's model: $(v_c)_{right} = (v_f)_{right} \cdot \exp\{-0.5\,[(k_c)_{right}/(k_c)_{right}]^2\}$, and thus the lane capacity $(c)_{right} = (v_c)_{right} \cdot (k_c)_{right}$.

The carriageway capacity can be estimated as the sum of the capacities of the three lanes. Similar consideration should be made for the critical density (k_c), while v_f and v_c can be respectively obtained from the traffic stream model (e.g. with Drake's model $v_f = c/(k_c \cdot \exp(-0.5))$ and the relation $v_c = c/k_c$.

6.4.2 *Case Study: Traffic Flow Parameters Estimation After HSR System Activation*

Table 6.3 shows the traffic flow parameters deduced for a two-lane divided highway. Each carriageway is composed of a travelling lane (right lane), an overtaking lane and an emergency lane. The aims are to assess the new traffic flow parameters, including the carriageway capacity, and to calibrate the Drake's traffic stream model, referred to HSR system activation.

Table 6.3 Initial traffic flow parameters (without HSR system)

Lane/carriageway	v_f (km/h)	k_c (pcu/km)	c (pcu/h)	v_c (km/h)
Travelling lane (right lane)	106	24	1552	65
Overtaking lane	128	25	1916	77
Carriageway	115	47	3254	70

In case of HSR system activation, from (6.18) and with the values shown in Table 6.3, it yields:

$$\frac{128 + (v_f)_{right}}{2} = 106 \rightarrow (V_f)_{right} = 84\,km/h$$

From (6.19) and with the values shown in Table 6.3, it follows:

$$\frac{25 + (k_c)_{right}}{2} = 24 \rightarrow (K_c)_{right} = 23\,puc/km$$

Through the Drake's expression (Table 2.2) the critical speed is obtained:

$$\begin{aligned}(v_c)_{right} &= (v_f)_{right} \cdot \exp\left\{-0.5\left[(k_c)_{right}/(k_c)_{right}\right]^2\right\} \\ &= 84 \cdot \exp\{-0.5[(23)/(23)]^2\} = 51\,km/h\end{aligned}$$

The capacity of the hard shoulder lane (i.e. "third dynamic lane") is:

$$(c)_{right} = (v_c)_{right} \cdot (k_c)_{right} = 51 \cdot 23 = 1173\,pcu/h$$

For the whole carriageway, the capacity $(c)_{carr}$, critical density $(k_c)_{carr}$, free-flow speed $(v_f)_{carr}$ and critical speed $(v_c)_{carr}$ are assessed as follows:

$$(c)_{carr} = (c)_{right} + (c)_{central} + (c)_{over} = 1173 + 1552 + 1916 = 4641\text{ pcu/h}$$
$$(k_c)_{carr} = (k_c)_{right} + (k_c)_{central} + (k_c)_{over} = 23 + 24 + 25 = 72\text{ pcu/km}$$
$$(v_f)_{carr} = c/[(k_c)_{carr} \cdot \exp(-0.5)] = 4641/[72 \cdot \exp(-0.5)] = 106\text{ km/h}$$
$$(v_c)_{carr} = (c)_{carr}/(k_c)_{carr} = 4641/72 = 64\text{ km/h}$$

The new traffic flow parameters are shown in Table 6.4, together with traffic flow laws $q = q(k)$.

Table 6.4 Traffic flow parameters after HSR system activation

Lane/carriageway	v_f (km/h)	k_c (pcu/km)	c (pcu/h)	v_c (km/h)	$q = q(k)$
Dynamic lane ("right")	84	23	1173	51	$84 \cdot k \cdot \exp[-0.5(k/23)^2]$
Travelling lane ("central")	106	24	1552	65	$106 \cdot k \cdot \exp[-0.5(k/24)^2]$
Overtaking lane ("over")	128	25	1916	77	$128 \cdot k \cdot \exp[-0.5(k/25)^2]$
Carriageway ("carr")	106	72	4641	64	$106 \cdot k \cdot \exp[-0.5(k/72)^2]$

6.5 Variable Speed Limits (VSL)

The hard-shoulder running system is generally associated to a VSL (*Variable Speed Limits*) system (Fig. 6.6).

VSL systems can have either a *mandatory rule* when they indicate users how to behave compulsorily or an *advisory rule* when they only suggest a speed limit with no obligatory compliance to it.

From a purely theoretical viewpoint, if almost all users are observant of imposed speed limits and the lane change frequency is very small, in order to obtain the traffic flow relations in highway segments with HSR and VSL systems, the vehicle space mean speed v is assumed to be constantly equal to the imposed limit value v_0. This means to set the following conditions between v and the other variables, flow q and density k:

$$v = v(k) = v_0 \tag{6.20}$$

$$v = v(q) = v_0 \tag{6.21}$$

By assuming $v = v_0 = \text{const.}$, from the fundamental flow relation $q = k \cdot v$, it follows:

Fig. 6.6 Highways with HRS and VSL systems

$$q = v_0 \cdot k \tag{6.22}$$

In order to adapt (6.20), (6.21) and (6.22) to lanes and the carriageway, the following conditions are to be taken into consideration [5, 13]:

- since the HSR system is activated only in case of high traffic demand values, Eqs. (6.20), (6.21) and (6.22) result to be non-significant for q values which can be associated to flow conditions varying from free to stable (Chap. 2);
- in a traffic regime tending to be homotachic (like that pursued with the speed limit imposition), traffic flows can reach the highest stability levels.

In these conditions, flows and the corresponding density values result to be both affected by very small dispersions. In case of VSL activation, critical density k_c values can be expected to be higher than those provided by a common traffic control (i.e. without VSL system).

In the absence of experimental determinations of capacity c, density k_c and critical speed v_c, in case of VSL systems with *mandatory rule,* the values assumed here for k_c are those in Table 6.4 ($k_c = 25$ pcu/km/lane for all lanes).

In addition, with Eq. (6.22), for given values of the speed limit v_0 the capacity c is calculated.

In short, the values in Table 6.5 and relations $q = q(k)$ and $v = v(q)$ in Fig. 6.7 are obtained.

For safety reasons, speed reduction must be imposed gradually to drivers, starting from the highest highway speed limit value (e.g. 130 km/h) up to the chosen v_0 value. The coordinated speed control system is based on real-time assessment of the reliability Φ and mean speed $\overline{v}_t$ in the control sections X_i (Fig. 6.8). In function of the observed values, the programme P_k, $k = 0, 1, \ldots, 4$ as well as the speed limits on the electronic traffic sign displays—which are around 1 km-distanced from each other—are selected. A VSL system control programme experimented on a German highway is illustrated in Fig. 6.8 [1–3, 5, 13].

Table 6.5 Traffic flow parameters for different speed limit values (v_0)

Lane/carriageway	v_0 (km/h)	k_c (pcu/km)	c (pcu/h)
Lane (indiscriminately: right/central/overtaking)	60	25	1500
	70	25	1750
	80	25	2000
Carriageway	60	75	4500
	70	75	5250
	80	75	6000

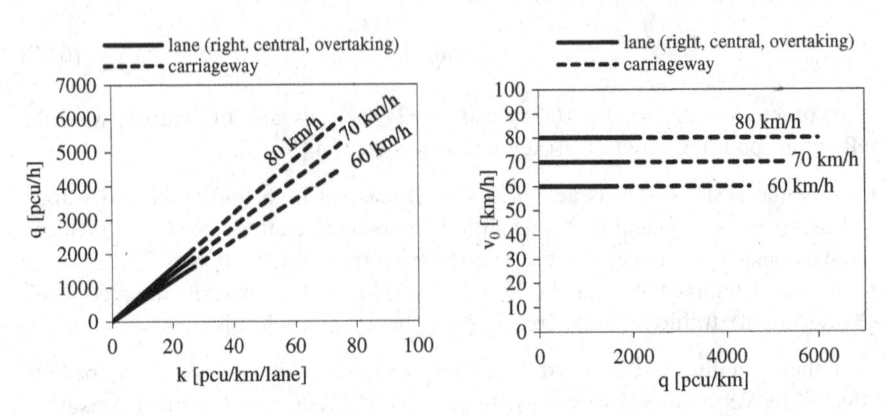

Fig. 6.7 Theoretical relationships q = q(k) and v = v(q) –VSL system with mandatory rule

V0 = (no limit sign)
V1 = 100
V2 = 80
V3 = 60
V4 = (warning sign)

SECTION	TRAFFIC CONDITIONS IN SECTION X_i				
	P0	P1	P2	P3	P4
	$\Phi \geq 0.98$	$0.98 < \Phi \leq 0.95$	$0.95 < \Phi \leq 0.90$	$30 < \overline{v}_t \leq 60$, $\Phi < 0.90$	$\overline{v}_t \leq 60$
X_{i-3}	V_0	V_0	V_0	V_1	V_2
X_{i-2}	V_0	V_0	V_1	V_2	V_3
X_{i-1}	V_0	V_1	V_2	V_3	V_4
X_i	V_0	V_1	V_2	V_3	V_4
X_{i+1}	V_0	V_1	V_2	V_3	V_3
X_{i+2}	V_0	V_0	V_1	V_2	V_2

X_{i-3} X_{i-2} X_{i-1} X_i X_{i+1} X_{i+2}

Fig. 6.8 Example of a VSL system control programme [1–3, 5, 13]

6.6 Automated Highway System (AHS)

The *Automated Highway System* (*AHS*) is an innovative highway transport system, at the experimental stage, characterised by traffic flow conditions markedly different from those typical of conventional highways. In fact, the traffic flow is composed of autonomous vehicles (AVs) and connected and autonomous vehicles (CAVs), travelling isolated or in platoons, and exchanging, mostly kinematic, information with the other vehicles and the infrastructure.

The first theoretical studies on self-driving vehicles date back to the 1940s, but only since the 1980s the applied research results have suggested the construction and installation, always experimentally, of the first working prototypes of automatically-driven vehicles [14, 15]. Thus, the projects ARAMIS, PROMETHEUS, VITA II, SARTRE, HAVEit were of great interest. These pioneering studies clearly revealed the advantages from the semi-automated and automated driving systems in terms of both functionality and safety, due to eliminating the human driver.

According to some researches [14, 16], the automated vehicles travelling in platoons—each composed of $N = 20$ vehicles and 60-m inter-distance between platoons, with speed of around 90 km/h—can determine a potential capacity up to 8000 veh/h per lane, while the usual lane capacity is of the order of 2000 veh/h (see Chap. 2).

Moreover, the reduced space gap (i.e. g_{si}, see Chap. 1) between vehicles of the same platoon (which can potentially be established at 1–2 m) implies, in the event of rear-end collision, reduced relative speeds between vehicles and therefore low impact energies, with potential benefits on safety [3].

Further benefits of the AHS system are environmental, thanks to minor fuel consumption (up to 25%) and air pollution emission reduction (up to 27%) [1] deriving from the decreased aerodynamic resistance to which platoon vehicles are subject (with the only exception of the leader vehicle).

In addition to the above potential benefits in terms of capacity, safety and energy efficiency and environmental sustainability, the implementation of AHS systems can also lead to significant economies for highway operators. Thanks to such a technology, some infrastructure upgrading (like, for instance, the construction of additional lanes traditionally required by the increasing traffic demand in the highway life-cycle) could be avoided.

6.6.1 *Estimation of the Increase in Lane Capacity*

Referring to Fig. 4.1 and recalling Eq. (1.12), the density k for a flow composed of vehicles of length l, with manual driving and speed v, can be expressed as [17, 18]:

$$k = \frac{1}{t_{pr} \cdot v + l} \tag{6.23}$$

The perception/reaction time t_{pr} varies with speed. For instance, Italian regulations [19] provide that $t_{pr} = 2.8–0.036 \cdot v$ (with v in m/s).

The capacity c_m of the lane travelled only by manually-driven vehicles is:

$$c_m = k_c \cdot v_c = \frac{v_c}{(2.8 - 0.036 \cdot v_c) + l} \tag{6.24}$$

where v_c is the critical speed and k_c is the critical density (Chap. 2).

For a lane travelled only by autonomous vehicles a relation analogous to (6.23) can be written only if considering a reaction delay t_a (i.e. the sum of sensing, computing, communication and brake actuation delays) as constant when v varies and much smaller than t_{pr}. Therefore, the capacity c_a of such a lane can be estimated as follows:

$$c_a = \frac{v_c}{t_a \cdot v_c + l} \tag{6.25}$$

The ratio χ between the two capacity values and hence the lane capacity increase is:

$$\chi = \frac{c_a}{c_m} = \frac{(2.8 - 0.036 \cdot v_c) + l}{t_a \cdot v_c + l} \tag{6.26}$$

For instance, for an overtaking lane, by introducing $v_c = 78$ km/h into (6.26) (see Table 2.3) and considering $t_a = 0.4$ s and $l = 7.5$ m (vehicle length 4.5 m + a safe 3 m distance between each vehicle pair), it follows $\chi = 3.2$. In other words, in the previously specified hypotheses, the overtaking lane capacity increases 3.2 times if it is used by only autonomous vehicles, compared to the homologous value of a conventional highway travelled by only manually-driven vehicles.

6.7 C-ITS, C-Roads and Smart-Roads

6.7.1 C-ITS

Cooperative Intelligent Transportation Systems are special types of ITS,[1] characterised by information and communication technologies which allow drivers and road operators to share real-time information aimed at improving functionality and safety in transport systems and user services.

C-ITS are based on three types of connectivity:

- *V2V* (vehicle to vehicle) *connectivity*: information exchange (e.g. speed and position) between two or more vehicles travelling on the same road;

[1] ITS (Intelligent Transportation Systems) are applied in a vast number of transport system fields and can be classified into [14]:

(1) ATMS (Advanced Traffic Management Systems): for advanced traffic management. They are systems for managing and controlling traffic, implemented especially in freeway segments and highway ramps (variable speed limits, ramp-metering installations, dynamic speed limits, etc.);
(2) ATIS (Traveler Advanced Information Systems): are on-board navigation/assistance systems for road users;
(3) AVCS (Advanced Vehicle Control Systems) for advanced vehicle control: collision warning systems, collision avoidance systems using automatic braking and/or automatic steering, autonomous vehicles travelling on dedicated highway lanes (the so-called Automated Highway System "AHS");
(4) CVO (Commercial Vehicle Operations): ITS technologies applied to commercial vehicles such as trucks, buses, vans, taxis and emergency vehicles;
(5) APTS (Advanced Public Transportation Systems): ATM, ATIS and AVCS implementation to improve the operation of high-occupancy vehicles (e.g. buses).

- *V2I* (vehicle to infrastructure) *connectivity*: information exchange between vehicle and road infrastructure;
- *I2V* (infrastructure to vehicle) *connectivity*: information from the road infrastructure to vehicle.

The connectivity between user, vehicle and road infrastructure is made by means of three different device typologies [14]:

- *Drivers' mobile device* through which they can receive information from road infrastructures and/or other vehicles;
- *On Board Unit* (OBU), devices which are installed on vehicles to transmit and receive information from and to the Road Side Unit (RSU);
- *Road Side Unit* (RSU), device fixed along the roadside, intersections and parking areas. An RSU both supports the Vehicle-to-Infrastructure (V2I) and increases the Vehicle-to-Vehicle (V2V) communication connectivity. It is used not only to receive and transmit information from and to vehicles, but also to process and send data to the traffic management control system in order to monitor traffic flow variables (flows, speeds, etc.).

C-ITS systems aim for the following key objectives:

- *Intelligent traffic management*: capacity increase, reduction in travel times and air pollutant emissions;
- *Mobility as a Service* (MaaS): integrated and "customised" transport service into a single mobility service, accessible on demand and based on individual needs.

Communication technology is based on *Wi-Fi in motion* allowing for the connectivity between drivers' mobile devices and road infrastructures, thus providing the regular information exchange from and to the user.

The transmission standards for Wi-Fi in motion, developed by IEEE1 802.11 working group, define the requisites concerning:

- *PHY* (*Physical Layer*), *OSI* (*Open System Interconnection*) level 1 which receives the bit-sequence of the exchanging message;
- *MAC* (*Media Access Control*), *OSI* level which acts as access control to PHY, in that it serves as interface with the client (connected road user) and the access point (device allowing for a Wi-Fi network connection).

6.7.2 C-Road Platform

On 30 November 2016, the European Commission approved a common strategy on intelligent cooperative transport systems (C-ITS) which has allowed testing innovative vehicles (i.e. autonomous vehicles) on European roads.

The C-ROAD platform was introduced to exchange experiences, projects and test results carried out in different European countries and to develop guidelines, specific techniques and harmonised standards (in accordance with the recommendations produced by the EU C-ITS platform) through coordination of both the implementation C-ITS projects run in European countries and C-ITS "Day-1" services.[2] On the whole, 16 pilot projects were launched [20].

6.7.3 Smart-Roads

Smart roads represent the new paradigm of mobility and aim at being sustainable, qualitatively innovative and inclusive, thus modifying the traditional relationship between vehicles, users and infrastructure/environment. It is *a digitalised infrastructure* increasingly compatible with the new technologies employed in light and heavy vehicles, with special regard to the more recent automated or semi-automated driving modes. In addition to the essential C-ITS (see Sect. 6.7.1), one or more of the following systems can be implemented in *smart roads*:

- *Lanes dedicated to autonomous vehicles* travelling isolated or in platoons (systems analogous to AHS, see Sect. 6.6);
- *Internet of Things* (*IOT*): sensors for monitoring traffic flows, structures (bridges, viaducts, road safety barriers etc.), weather and air pollutants;
- *Ramp-metering systems* (see Sect. 6.3);
- *HSR systems* (see Sect. 6.4);
- *VLS systems* (see Sect. 6.5);
- *Green Island*: a multi-technological site (e.g. one for each 30 km highway segment) for generating energy from renewable sources, equipped with solar photovoltaic cells, mini-wind turbines, etc. It allows for a highway segment to be power-supplied, thus lowering operating costs. Every Green Island can be provided with power recharge points, *drone zones for landing and take-off* used to monitor highways and deliver the first-aid kit like, for instance, portable defibrillators;
- *Electric priority lanes* (lanes only for electric vehicles equipped with wireless recharge technology).

[2]C-ITS Day-1 services include:

(1) *Hazardous location notification* (warning of a slow or stationary vehicle or when approaching traffic jam; roadworks warning; weather conditions; emergency electronic brake light; approaching emergency vehicle; other risks);

(2) *Signage services* (on-board signals; in-vehicle speed limits; signal violation; intersection safety; traffic signal priority request by specific vehicles; green light optimal speed advisory; cooperative vehicle data; shockwave damping.

6.8 Crash Frequency Estimation: The HSM Method

The annual "*crash frequency*[3]" expected on roads and intersections can be estimated with the Highway Safety Manual (HSM).

The HSM provides analytical tools and techniques for quantifying the potential effects on crashes as a result of decisions made in planning, design, operations and maintenance [21].

The predictive accident models are distinctively based on the road type to analyse: Rural Two-Lane Roads (segments and intersections); Rural Multilane Highways (segments and intersections); Urban and Suburban Arterials (segments and intersections).

In particular, in highways the predictive method applies to the following freeway facilities [21]:

- rural freeway segment with four to eight lanes;
- urban freeway segment with four to ten lanes;
- freeway speed-change lanes associated with entrance ramps and exit ramps.

For freeway segments, the procedure can be applied in case of annual average daily traffic (AADT) threshold levels given in Table 6.6.

The HSM application to sites with traffic volumes substantially outside the ranges in Table 6.6 may be little reliable.

The predicted average crash frequency N_T of a freeway segment is obtained as the sum of the following four terms:

- N_1 = predicted average crash frequency of a freeway segment with multiple vehicles involved in fatal or injury crash;
- N_2 = predicted average crash frequency of a freeway segment with single vehicle involved in fatal or injury crash;
- N_3 = predicted average crash frequency of a freeway segment with multiple vehicles involved, and property damage only;

Table 6.6 Maximum AADT values for which HSM relations are reliable

Field	Cross section (through lanes)	AADT (veh/day)
Rural	4	0–73,000
	6	0–130,000
	8	0–190,000
Urban (more than 5000 inhabitants)	4	0–110,000
	6	0–180,000
	8	0–270,000
	10	0–310,000

[3]Number of crashes occurring at a particular site, facility or network in a one-year period (Crash frequency = Number of crashes/Period in years).

- N_4 = predicted average crash frequency of a freeway segment with single vehicle involved, and property damage only.

Thus it follows:

$$N_T = \sum_{i=1}^{4} N_i \quad (6.27)$$

Each of the previous crash frequencies N_i (with i = 1, 2, 3 and 4) is calculated with an expression of the type:

$$N_i = N_0 \times (CMF_1 \times CMF_2 \times \cdots \times CMF_{11}) \times C_i \quad (6.28)$$

where

- N_0 (or SPF) *safety performance function* = predicted average crash frequency of a freeway segment with base conditions,[4] expressed in crashes/year;
- CMF_j (with j = 1, 2 … 11) *crash modification factors* = multiplicative factors to modify the predicted average crash frequency in base conditions (N_0) in order to take into account the real (geometric and traffic) characteristics of the road under study;
- C_i = *calibration factor* to adjust the *safety performance function* (N_0) to local conditions. The *calibration factors* depend on the examined territorial context (e.g. country) and therefore on the accident time series data of highways similar to that under study.

The *safety performance function* has the following expression:

$$N_0 = L \times e^{a+b\times\ln(c\times AADT)} \quad (6.29)$$

where L denotes the effective length of a freeway segment (expressed in miles), AADT the relative annual average daily traffic, a and b the regression coefficients (to be determine according to the guidelines in the same manual [21]) and c the AADT scale coefficient.

Finally, the *calibration factor* C_i derives from the ratio between the total crash number observed (i.e. those effectively occurred in the past) in a chosen set of

---

[4]For example, the main base conditions for multiple-vehicle crashes are:

- length of horizontal curve = 0 m (i.e., not present);
- lane width = 3.6 m; inside shoulder width (paved) = 1.8 m; median width = 18 m;
- length of median barrier = 0 m (i.e., not present);
- number of hours where volume exceeds 1000 veh/h/lane = none;
- distance to nearest upstream ramp entrances >800 m;
- distance to nearest downstream ramp exits >800 m.

highway sites[5] similar to that under study, and the total predicted crash number assessed with the model. Therefore, the value of C_i can be less than, equal to, or greater than 1.

6.9 Models for Estimating Traffic Pollutant Emissions

Traffic emissions depend on numerous factors, such as vehicle types (light, heavy), fuel characteristics (gasoline, diesel, liquefied petroleum gas LPG, etc.), vehicle operating conditions (speed, driving style, acceleration, deceleration, etc.), infrastructure types (urban, rural, intersections, etc.).

Vehicle emission models are developed in order to assess the impact of traffic flows on air quality. In analogy to traffic models (see Sect. 4.6), also emission models can be classified as follows [22]:

- *macroscopic models*, based on aggregate variables, which are used to evaluate overall pollutant emissions in a road segment;
- *microscopic models*, used to estimate the emission of each vehicle in detail, starting from an accurate description of its kinematic variables over the time (speed, acceleration, deceleration, etc.);
- *mesoscopic models*, differing from the previous two models in having an intermediate level of detail.

Below the macroscopic model implemented in COPERT software (Computer Programme to calculate Emissions from Road Traffic Traffic) will be analysed in detail.

6.9.1 The Macroscopic Model COPERT

The software was developed by the European Environment Agency (EEA) inside the CORINAIR programme [23]. COPERT is an aggregate model. It allows to obtain emission values for every vehicle category. It particularly allows for estimating the following main pollutants CO (carbon monoxide), NO_x (nitrogen oxides), SO_x (sulphur oxides), VOC (volatile organic compounds), N_2O (nitrous oxide), NH_3 (ammonia), CO_2 (carbon dioxide), CH_4 (methane), Pb (lead), $PM_{2,5}$ and PM_{10} (particulate matter), NMVOC (non-methane volatile organic compound), apart from fuel consumption.

According to this model, traffic emissions depend on [23]:

- (gasoline, diesel, catalysed and non-catalysed vehicles), vehicle mass (cars, heavy vehicles, motorcycles);

[5]The desirable minimum sample size for the calibration database for one predictive model is 30–50 sites. For segments, each site should be between 0.1 and 1.0 min in length [21].

- travelling type (urban, extra-urban and rural);
- travelling length;
- weather conditions;
- road longitudinal gradient;
- traffic space mean speed;
- etc.

The model associates the functions for estimating speed-based emission and fuel consumption to every vehicle type. These functions represent the mean curves of emission and fuel consumption inferred from data empirically observed in different European countries. The emissions E_i, of the pollutant "i", are calculated by the sum of three distinct components:

$$E_i = E_{hot,i} + E_{cold,i} + E_{vap,i} \tag{6.30}$$

where

- $E_{hot,i}$: hot emissions, produced in the stabilised engine operation (after the catalyst warm-up phases are completed). The value of these emissions are related to travelled distance, vehicle type, speed, age, engine type (legislation on emission category: Euro I, …, Euro VI) and weight;
- $E_{cold,i}$: cold emissions, produced during the engine warm-up phase (CH_4 and CO emissions are elevated given that the catalyst must reach the operating temperature);
- $E_{vap,i}$: evaporative emissions consisting only of NMVOCs (Non-Methane Volatile Organic Compounds), associated with the evaporative phenomenon.

In general, total emissions E of a long distance trip can be evaluated, as follows:

$$E = E_{urban} + E_{rural} + E_{highway} \tag{6.31}$$

in which E_{urban}, E_{rural}, $E_{highway}$ are the emissions in urban (v = 10–50 km/h), rural (v = 40–80 km/h) and highway (v = 70–130 km/h) contexts.

In order to estimate hot emissions, the following expression is used [23]:

$$E_{hot,i} = \sum_{jk} n_j \cdot m_{jk} \cdot e_{hot;ijk} \; [t/year] \tag{6.32}$$

with:

- n_j vehicle number, class j-th (vehicles);
- m_{jk} annual average distance travelled by every category j vehicle on a k-class road (km/vehicle);
- $e_{hot;ijk}$ emission factor for the pollutant i, observed in category j vehicles on a k-class road (t/km).

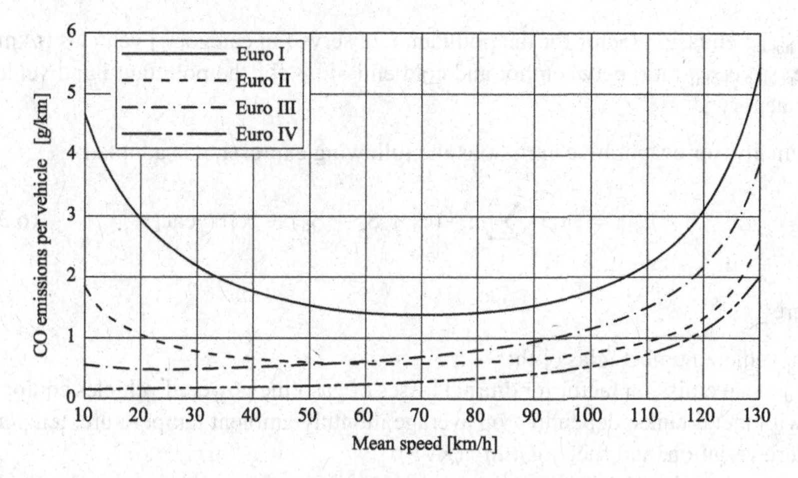

Fig. 6.9 CO emissions in function of the mean speed—COPERT model [22]

The emission factor depends on the vehicle speed. By way of an example, Fig. 6.9 shows the curves $e_{hot;ijk} = e_{hot;ijk}(v)$ for the pollutant CO of different vehicle types (from Euro I to Euro IV). Clearly, in old vehicles (i.e. legislation on emission category Euro I and Euro II) emissions have low values in an intermediate speed range. On the other hand, in more modern vehicles (Euro III and Euro IV) emissions increase monotonically with speed.

In order to assess cold emissions, the following expression is used, in that it takes into account the *emission surplus* which would be present if all vehicles always worked at operating temperature:

$$E_{cold,i} = B - (A \cdot E_{hot,i}) \tag{6.33}$$

where

- B denotes the emissions produced until the engine reaches the operating temperature;
- A stands for the multiplicative factor in order not to count twice the hot emissions occurring in this phase.

The expression to determine cold emissions is the following:

$$E_{cold,i} = \sum_{jk} \beta_j \cdot n_j \cdot m_j \cdot e_{hot;i,j} \cdot \left(\frac{e_{cold;i,j}}{e_{hot;i,j}} - 1 \right) \text{[t/year]} \tag{6.34}$$

In which.

- β_j: km fraction travelled with the cold engine (or catalyst) for category j vehicles;
- n_j: vehicle number, class j-th (vehicles);
- m_j: annual average distance travelled by every category j vehicle (km/vehicle);

- $e_{hot;i,j}$: emission factor for the pollutant i, observed in category j vehicles (t/km);
- $e_{cold;ij}/e_{hot;ij}$ ratio between hot and cold emissions for the pollutant i and vehicle category j.

Finally, for evaporative emissions the following expression is adopted:

$$E_{vap} = 365 \cdot \sum_{j} n_j \cdot (e_d + S_c + S_{fi}) + R\,[t/year] \tag{6.35}$$

where

- n_j vehicle number, class j-th;
- e_d mean emission factor for diurnal losses of gasoline powered vehicles equipped with metal tanks, depending on average monthly ambient temperature, temperature variation, and fuel volatility (RVP);
- S_c average hot and warm soak emission factor of gasoline powered vehicles equipped with carburettor;
- S_{fi} average hot and warm soak emission factor of gasoline powered vehicles equipped with fuel injection;
- R hot and warm running losses.

Clearly enough, a mean speed reduction on a highway, e.g. thanks to VSL systems (Sect. 6.5), can lead to a consequent reduction in pollutant emissions (see Fig. 6.9).

By way of an example, consider the case of Austrian road operator ASFINAG who implemented a VSL system (from 130 to 100 km/h) on 5 highway segments. Thanks to this system, the following benefits derived [24]: reduction in NO_x emissions between 3.8 and 10.1%, and in CO_2 between 2.6 and 6.7%. NO_2 emissions reduced by 2.5–5.6%.

References

1. Ferrari P (2007) Theory and control of traffic flow. TEP-Tipografia Editrice Pisana (in Italian)
2. Ferrari P (1988) The reliability of motorway transport system. Transp Res Part B 22(4):291–310
3. Ferrari P (1991) The control of motorway reliability. Transp Res Part A 25(6):419–427
4. Vitetta A (2003) Traffic flows in transport systems. Franco Angeli (in Italian)
5. Brilon W (2012) Reliability of motorway operations. In: Conference "traffic safety policies: an international comparison of policy changes". International Association of Traffic and Safety Sciences. Tokyo, 20 Sept 2012
6. Papageorgiou M (1991) Concise encyclopedia of traffic and transportation. Pergamon Press
7. Mauro R, Giuffrè O, Granà A (2013) Speed stochastic processes and freeway reliability estimation: evidence from the A22 freeway, Italy. J Transp Eng 139:1244–1256
8. Papageorgiou M, Kotsialos A (2002) Freeway ramp metering: an overview. IEEE Trans Intell Transp Syst 3(4):271–281
9. Papageorgiou M, Blosseville JM, Hadj-Salem H (1998) La fluidification des rocades de l'Ile de France: un projet d'importance. Dynamic Systems and Simulation Laboratory. Technical University of Crete, Chania, Greece, Internal Report 1998–17
10. Elefteriadou L (2014) An introduction to traffic flow. Springer

11. Brinckerhoff P (2010) Synthesis of active traffic management experiences in Europe and the United States, FHWA-HOP-10-031
12. Mauro R (2006) Variable Speed Limits and hard-shoulder running systems: traffic control analysis for the A22 motorway. Internal Report (available by R Mauro: raffaele.mauro@unitn.it)
13. Guerrieri M, Mauro R (2016) Capacity and safety analysis of hard-shoulder running (HSR). A motorway case study. Transp Res Part A 92:162–183
14. Ioannou A et al (1997) Automated highway systems. Springer Science+Business Media, LLC
15. Carbaugh J, Godbole DN, Sengupta R (1998) Safety and capacity analysis of automated and manual highway systems. Transp Res Part C Emerg Technol 6:69–99
16. Horowitz R, Varaiya P (2000) Control design of an automated highway system. IEEE Proc IEEE 88(7):913–925
17. Mauro R, Guerrieri M (2017) Functional and safety analysis of hard-shoulder running system (HSR). The case study of the A22 motorway. Internal Report (available by R Mauro: raffaele.mauro@unitn.it)
18. Friedrich B (2016) The effect of autonomous vehicles on traffic. In: Maurer M, Gerdes J, Lenz B, Winner H (eds) Autonomous driving. Springer
19. Italian Guidelines for the Design of Roads and Highways (D.M. 5/11/2001).
20. C-roads pilot overview (report version 1.0), Dec 2017. Available https://www.c-roads.eu
21. Highway Safety Manual (HSM) (2014) American Association of State Highway and Transportation Officials, AASHTO
22. Ferrara A, Sacone S, Siri S (2018) Freeway traffic modelling and control. Springer
23. https://emisia.com/products/copert
24. Thudium J, Chelala C, Immissionsgesteuerte Geschwindigkeitsbegrenzung auf Österreichs Autobahnen. Report available: www.oekoscience.ch/Bibliothek/Lufthygiene/Oekoscience_ImmissionssteuerungFarbig.pdf

Chapter 7
Interference Between Traffic Flows: The Gap Acceptance Theory

Abstract This chapter presents the gap acceptance theory, in that it is of great applicative interest in studying road intersections. Therefore, some models for estimating the critical gap and the follow-up time for at-grade unsignalized intersections and roundabouts are described.

The traffic flow models described in the previous chapters can be applied when there are no mutual interferences between traffic flows.

In some segments of the road network (weaving areas, deceleration and entry lanes) and at junctions (at-grade intersections), vehicle flows interfere with one another. Consequently, driver behaviors need to be properly modeled, also in function of systems of traffic flow regulation (priority rules). A modeling carried out correctly makes it possible to obtain reliable estimations of capacity, delays and queues.

According to the *gap acceptance theory,* at unsignalized at-grade intersections, drivers who have to cross or enter a traffic stream (of a given flow) perform the following logical and behavioral processes:

- when they arrive at the yield or stop lines, they slow their motion down and start evaluating the time headways (or *gaps*) between the vehicle pairs following each other in the main stream to be crossed or entered;
- their estimations are not distorted, that is every perceived gap corresponds to the real time headway h_{ti} between vehicles;
- drivers make a systematic comparison between every vehicle headway h_{ti} and their own *critical gap* T_c;
- the critical gap T_c is the smallest time interval between the vehicles in the major-street (main traffic stream), thought as acceptable to make the desired maneuver in safety conditions (see Sect. 7.2);
- a driver can make the maneuver when the gap is higher than or equal to the own critical gap ($h_{ti} \geq T_c$). The maneuver is performed as soon as the first suitable gap opens up in the traffic stream so as to satisfy the aforesaid condition.

M. Guerrieri and R. Mauro, *A Concise Introduction to Traffic Engineering*,
Springer Tracts in Civil Engineering,
https://doi.org/10.1007/978-3-030-60723-4_7

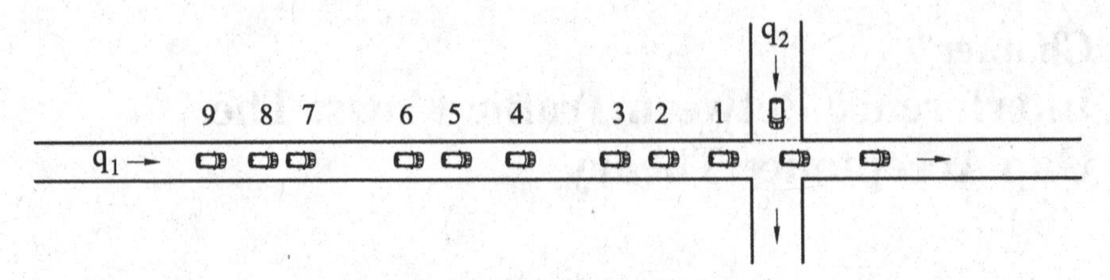

Fig. 7.1 Example of an at-grade intersection with a vehicle waiting at the stop line

Table 7.1 Comparison between time headways in the main flow and drivers' critical gaps in the secondary flow (see Fig. 7.1)

Vehicle pairs (main flow q_1)	Time headways h_{ti} (s)	Comparison between h_{ti} and $T_c = 6$ s	Crossing possibility of flow q_1
1–2	3.1	$h_{ti} < T_c$	No
2–3	2.8	$h_{ti} < T_c$	No
3–4	7.8	$h_{ti} > T_c$	Yes
4–5	2.5	$h_{ti} < T_c$	No
5–6	2.4	$h_{ti} < T_c$	No
6–7	8.1	$h_{ti} > T_c$	Yes
7–8	1.9	$h_{ti} < T_c$	No
8–9	2.1	$h_{ti} < T_c$	No

In order to clarify these logical and behavioral processes, consider the intersection in Fig. 7.1. The vehicle of the secondary flow q_2, waiting at the stop line, must cross the intersection. The vehicle has to cross the main flow q_1, where vehicle gaps have a time length h_{ti} (see Table 7.1). Assume that the driver under examination has a critical gap $T_c = 6$ s.

The first suitable time interval higher than T_c ($h_{ti} \geq T_c$) appears between vehicles Nos. 3 and 4, which are $h_{ti} = 7.8$ s away from each other.

Therefore, the vehicle under examination passes the flow q_2 just across vehicles Nos. 3 and 4.

On the other hand, the driver cannot perform any maneuver before vehicle No. 3 passes by since the gaps between vehicles Nos. 1 and 2 and between vehicles Nos. 2 and 3 are lower than T_c ($h_{ti} < T_c$).

Clearly, the drivers in a secondary flow q_2 who will be able to cross or enter the main flow q_1 (which has a priority over q_2), in a given time headway, are going to increase when both the critical gap T_c and the flow q_1 decrease.

In fact, for a given value T_c, when q_1 decreases, according to Eq. (1.6), the average headway between vehicles in the main stream increases; this leads to an increase in entries from the secondary street. Analogously, once q_1 is set, the number of entries from the secondary street grows when T_c diminishes, since there is an increase in cases with $h_{ti} > T_c$.

Sometimes more vehicles coming from the secondary flow can turn into the same gap, provided it has a sufficiently high length. Should this be the case, the time headway between vehicle pairs of the secondary flow q_2 exploiting a single gap in the main flow q_1, is called *follow-up time* T_f (measurable by in situ traffic surveys).

According to the *gap acceptance theory*, the capacity c associated to a crossing or entry stream from a secondary to a main street or from a secondary to a main lane (e.g. highway entry lanes) is a function of type:

$$c = c(q_1, T_c, T_f) \tag{7.1}$$

The critical gap T_c and the follow-up time T_f are the so-called *drivers' psychotechnical parameters.*

T_c and T_f depend on the type and complexity of maneuvers and from behavioral attitudes of driver population.

In order to analyze the performance indicators of intersections (and other types of road infrastructures with conflicting streams) it is essential to reach reliable estimations of T_c and T_f for each maneuver.

7.1 Estimation of the Critical Gap and Follow-Up Time

The example provided in the previous section shows that the time headways between vehicles in the main flow which are rejected by secondary street drivers can be measured.

In other words, the length of the time headways can be measured when the condition $h_{ti} < T_c$ occurs. Also the gap (resulting in $h_{ti} > T_c$) used to perform the maneuver can be measured.

In fact, always referring to the example in Fig. 7.1 and Table 7.1, an external observer can only infer that:

- the critical gap T_c of the driver in the secondary flow is higher than the greatest of the rejected time headways ($T_c > 3.1$ s);
- the critical gap T_c of the driver is lower than or equal to the time headway used to perform the crossing maneuver $T_c \leq 7.8$ s.

Therefore, for the driver previously considered, it is possible to determine only the interval of values encompassing T_c. Thus it follows: $3.1 \text{ s} < T_c \leq 7.8$ s.

For every vehicle in the secondary flow an analogous value interval can be deduced.

Consequently, the exact value T_c cannot be measured but only estimated.

The methods used for estimating the critical gap T_c are based on the following hypotheses:

- T_c is heterogeneous: the critical gap is variable in the driver population;
- T_c is a random variable with unknown probability law;

- T_c is constant for each driver (namely, it does not change over time).

The following methods are among the most used for estimating the average critical gap of the driver population $\overline{T}_c$ [1, 2]:

- Drew's method;
- the step method;
- Raff's method;
- Ashworth's method;
- Miller's method.

Generally speaking, in performance analysis the critical gap of all the drivers performing a given maneuver is assumed to be constant and with a value equal to the average critical gap of the driver sample under examination, that is: $T_c = \overline{T}_c$.

7.1.1 Estimation of the Average Critical Gap: Ashworth's Method

Ashworth's method [3, 4] can be applied if the vehicle headways in the main stream with flow q_1 are distributed according to a negative exponential law (see Chap. 5) and the accepted gaps T_a and the critical ones T_c of the drivers in the secondary stream follow a normal distribution.

Under these hypotheses, the average critical gap $\overline{T}_c$ is given by the relation [3, 4]:

$$\overline{T}_c = \overline{T}_a - q_1 \cdot \sigma_a^2 \tag{7.2}$$

Being $\overline{T}_a$ and σ_a^2 respectively the mean value and variance of the gaps accepted by the driver sample under examination.

In the previous expression, $\overline{T}_a$ and $\overline{T}_c$ are expressed in seconds and q_1 in vehicles per second.

7.1.2 Estimation of the Average Critical Gap: Miller's Method

The method is based on the hypothesis that the critical gaps T_c have a Gamma distribution. The average critical gap $\overline{T}_c$ can be obtained by solving the system [2, 5]:

$$\begin{cases} \overline{T}_c = \overline{T}_a - q_1 \cdot \sigma_c^2 \\ \sigma_c^2 = \sigma_a^2 \cdot \dfrac{\overline{T}_c}{\overline{T}_a} \end{cases} \tag{7.3}$$

Since q_1 denotes the flow of the main stream, $\overline{T}_a$ the average value of the gaps accepted by the drivers in the secondary stream, σ_a^2 and, σ_c^2 the variance of the accepted gaps and that of the critical gaps respectively. By means of the relations (7.3) the values of $\overline{T}_c$ and σ_c^2 can be inferred after calculating $\overline{T}_a$ and σ_a^2 on the basis of a proper value sample of accepted gaps T_a.

7.2 Characteristic Values of the Critical Gap and the Follow-Up Time

If estimations based on empirical data analysis (ad hoc measurements) cannot be carried out, it is then possible for T_c and T_f to be attributed values drawn from the literature as specified below.

7.2.1 T_c and T_f for Unsignalized Intersections

According to the Highway Capacity Manual, the critical gap T_{cx} and the follow-up headway T_{fx} concerning each minor movement "x" (crossing, right-turning, left-turning, etc.) can be computed separately with the following expressions:

$$T_{c,x} = T_{cb} + t_{c,HV} \cdot P_{HV} + t_{cG} \cdot G - t_{3,LT} \quad (7.4)$$

$$T_{f,x} = T_{fb} + t_{f,HV} \cdot P_{HV} \quad (7.5)$$

being:

- T_{cb} the base critical gap associated to a given maneuver (see Table 7.2);
- T_{fb} the basic follow-up time associated to a given maneuver (see Table 7.2);

Table 7.2 Base critical gaps T_{cb} and follow-up times T_{fb} at unsignalized intersections

Maneuver	T_{cb} (s)		T_{fb} (s)
	Two-lane major street	Four-lane major street	
Left-turn from the major street	4.1	4.1	2.2
Right-turn from the minor street	6.2	6.9	3.3
Through traffic on minor street	6.5	6.5	4.0
Left-turn from minor street	7.1	7.5	3.5

- $t_{c,HV}$ adjustment factor for heavy vehicles (1.0 for two-lane major streets and 2.0 for four-lane major streets) (s);
- P_{HV} the percentage of heavy vehicles in the secondary stream;
- t_{cG} the corrective coefficient due to the effect of the longitudinal slope, equal to 0.1 s for right-turning maneuvers from secondary to main street, and 0.2 s for crossing maneuvers from the secondary street and left-turning maneuvers from secondary to main streets;
- G longitudinal slope expressed in percentage;
- $t_{3,LT}$ adjustment factor for intersection geometry, equal to 0.7 s for left-turning maneuvers from the secondary street at three-leg intersections; 0 s in the other cases.

The basic values (T_{cb} and T_{fb}) which appear in-expressions (7.4) and (7.5) are given in Table 7.2.

7.2.2 T_c and T_f for Roundabouts

The reference minimum and maximum values used for conventional roundabouts of type "*single lane*" (a lane at entries and on the ring), "*double lane*" (two lanes at entries and two ring lanes) and for turbo-roundabouts (see Chap. 8) are shown in Table 7.3 [6, 7].

Table 7.3 Critical gaps and follow-up times for roundabouts

Geometric layout	T_c (s)		T_f (s)	
	Min.	Max.	Min.	Max.
Single-lane	3.16	5.50	2.10	4.0
Double-lane	2.03	5.50	1.60	5.00
Turbo-roundabout	4.03	5.48	2.52	2.71

7.3 The Theoretical Capacity of Traffic Streams in an Unsignalized At-Grade Intersection

Consider, once again, a traffic stream of flow q_2 whose motion is hindered by the main stream q_1 (see Fig. 7.1). Both flows are in stationary conditions. Moreover, be:

- T_c and T_f the psychotechnical parameters of the drivers in the secondary stream, with a flow q_2;
- $f(\tau)$ the probability density function of vehicle headways of the main stream (q_1), here denoted with τ (instead of h_{ti} as done previously, see Chap. 1) for the sake of simplicity;
- $n(\tau)$ the number of secondary stream vehicles which perform the desired maneuver entering the time gap of length τ;
- $\bar{n}$ the average number value of the secondary stream vehicles which enter the gaps in the main flow, completing the desired maneuver.

Given the hypothesis above, it follows:

$$\bar{n} = \int_0^{\infty} n(\tau) \cdot f(\tau) \cdot d\tau \tag{7.6}$$

The *capacity* of the entry or crossing maneuvers from secondary to main streets is given by the ratio between the average number of secondary stream vehicles performing the desired maneuver $\bar{n}$ and the average headway between the vehicles in the main stream $\bar{\tau} = 1/q_1$ (see Eq. 1.6):

$$c = \frac{\bar{n}}{\bar{\tau}} = q_1 \cdot \int_0^{\infty} n(\tau) \cdot f(\tau) \cdot d\tau \tag{7.7}$$

Assume the following behavioral model (Fig. 7.2) [8]:

- 0 vehicles perform the maneuver in the presence of a gap $\tau < T_c$;
- 1 vehicle performs the maneuver if $T_c \leq \tau < T_c + T_f$;
- 2 vehicles perform the maneuver if $T_c + T_f \leq \tau < T_c + 2T_f$;
- etc.

The above model makes it possible to obtain the step law $n = n(\tau)$ represented in Fig. 7.2, which can be approximated to the straight line (continuous function) with an intercept T_0, slope coefficient $1/T_f$ and through which it follows: $T_c = T_0 + 0.5 \cdot T_f$.

The resulting straight line is an example of the so-called "*gap acceptance model of continuous type*" [9], expressed by the following equation:

$$n(\tau) = \frac{\tau - T_0}{T_f} \tag{7.8}$$

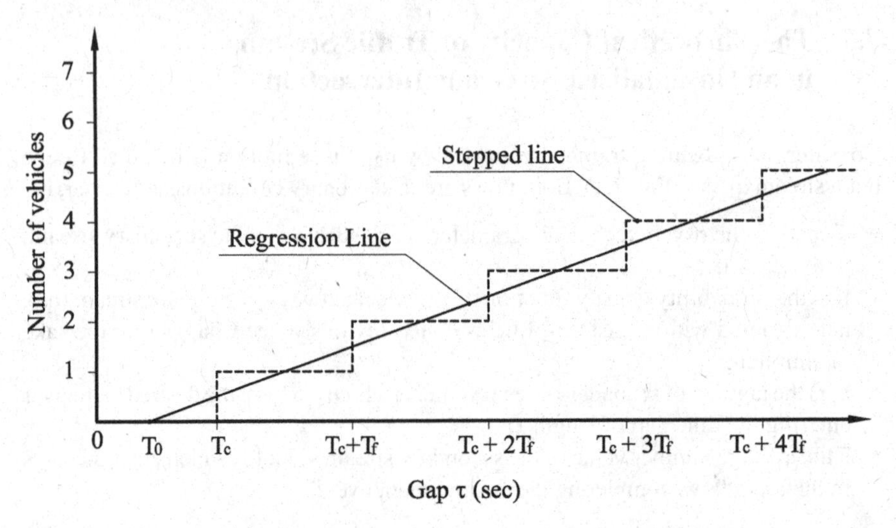

Fig. 7.2 Number of vehicles performing the maneuver in function of the τ value

For the vehicle headways τ in the main flow, consider an exponential law (see Chap. 5):

$$f(\tau) = q_1 \cdot e^{-q_1 \cdot \tau} \tag{7.9}$$

By inserting Eqs. (7.8) and (7.9) into (7.7) and solving them, it yields:

$$c = q_1 \cdot \int_0^\infty \frac{\tau - T_0}{T_f} \cdot q_1 \cdot e^{-q_1 \cdot \tau} \cdot d\tau = \frac{1}{T_f} \cdot e^{-q_1 \cdot \tau} \tag{7.10}$$

On the other hand, by considering the step law, the following expression results after some passages [8]:

$$c = \frac{q_1 \cdot e^{-q_1 \cdot T_c}}{1 - e^{-q_1 \cdot T_f}} \tag{7.11}$$

The results obtained from Expressions (7.10) and (7.11) substantially coincide if T_c, T_0 and T_f satisfy the relation $T_c = T_0 + 0.5 \cdot T_f$.

The expression (7.11) above is at the basis of the calculation models in the Highway Capacity Manual (HCM 2016) [10].

More in detail, at an at-grade intersection, the potential capacity $c_{p,x}$ of a given maneuver x, is calculated by the relation [9]:

$$c_{p,x} = q_{c,x} \cdot \frac{e^{-q_{c,x} \cdot \frac{T_{c,x}}{3600}}}{1 - e^{-q_{c,x} \cdot \frac{T_f}{3600}}} \tag{7.12}$$

Being $q_{c,x}$ the conflicting traffic flow (expressed in veh/h); $q_{c,x}$ is calculated by a procedure specified in the HCM 2016. $q_{c,x}$ depends on the maneuver type (crossing, right- or left-turning) and on the hierarchy attributed to the different traffic streams interfering with the maneuver x.

References

1. Esposito T, Mauro R (2003) Fundamentals of highway engineering, vol 2. Hevelius edizioni (In Italian, available by info@hevelius.it)
2. Brilon W, Koenig R, Troutbeck RJ (1999) Useful estimation procedures for critical gaps. Transp Res Part A 33:161–186
3. Ashworth R (1968) A note on the selection of gap acceptance criteria for traffic simulation studies. Transp Res 2(2):171–175
4. Ashworth R (1970) The analysis and interpretation of gap acceptance data. Transp Sci 4(3):270–280
5. Miller AJ (1974) A note on the analysis of gap—acceptance in traffic. J Roy Stat Soc 23(1):66–73
6. Giuffrè O, Granà A, Tumminello ML (2016) Gap-acceptance parameters for roundabouts: a systematic review. Eur Transp Res Rev 8(1):1–20
7. Guerrieri M, Mauro R, Parla G, Tollazzi T (2018) Analysis of kinematic parameters and driver behavior at turbo roundabouts. J Transp Eng Part A Syst 144(6), Art. n. 04018020
8. Papageorgiou M (1991) Concise encyclopedia of traffic and transportation. Pergamon Press
9. Plank AW (1982) The capacity of a priority intersection—two approaches. Traffic Eng Control 23(2):88–92
10. HCM (2016) Highway Capacity Manual: HCM 2016. Transportation Research Board, Washington, D.C.

Chapter 8
Queue Formation: General Models

Abstract This chapter deals with the models for studying queues in road transport systems under a unitary approach. They are the probabilistic models for stationary state, the deterministic solutions in congestion, the heuristic solutions for stationary and non-stationary states. The treatment of these models is unusual but can be directly applied to practical cases.

The formation of vehicle queues affects specific types of interrupted flow highway facilities, such as: road junctions, ramps, weaving zones and freeway toll booths (see Chap. 1). On these highway facilities, traffic streams converge and diverge and often stop for priority rules between them. Queues also occur, however, for example on motorways, which have uninterrupted flow operations, in the presence of peak traffic demand.

The estimation of queue length and waiting times is therefore of considerable applicative interest in highway engineering, to provide geometric design criteria and analysis of operating conditions.

The results of mathematical queue theory are used to study traffic waiting phenomena [1, 2].

The evolution over time of a queue of vehicles that have to perform a maneuver is connected to two distinct random processes:

- the arrival process (see Chap. 5), which represents the evolution of service demand (in other words, the amount of incoming users who want to perform the maneuver);
- the service process, which represents the evolution over time of the number of users served (i.e. the number of users performing the maneuver).

Typically, these two processes are independent of each other and their interactions within the waiting system are studied using queue theory models.

Figure 8.1 shows two typical cases of application of queue (or waiting) models in highway engineering: a simple road intersection and a freeway tollbooth.

The author of this Chapter is Andrea Pompigna.

M. Guerrieri and R. Mauro, *A Concise Introduction to Traffic Engineering*,
Springer Tracts in Civil Engineering,
https://doi.org/10.1007/978-3-030-60723-4_8

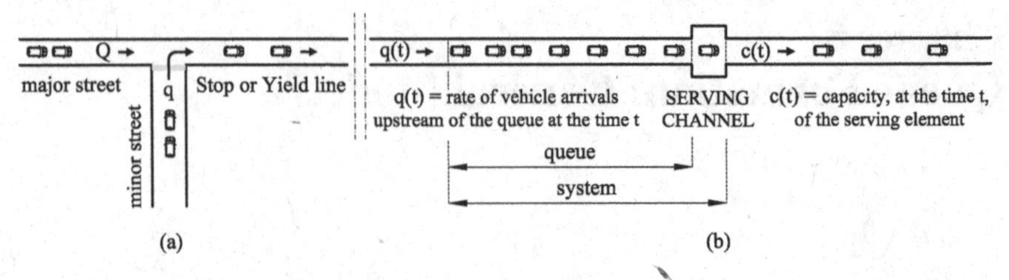

Fig. 8.1 Examples of simple applications of queueing systems: **a** road intersection, **b** freeway tollbooth

In the simple case of the intersection in Fig. 8.1a the queue emerges on the minor road and affects the flow q (arrivals) that wants to turn right to merge the flow Q on the main road. The queue originates because the major stream Q has priority over the minor stream q.

In the simple case of the freeway tollbooth in Fig. 8.1b the queue forms on the lane leading to the tollbooth and affects the vehicles of flow q whose users have to pay the toll (arrivals). The number of vehicles whose users pay the toll (vehicles served) depends on the duration of the payment transactions. The queue arises because of the time interval in which each user stops to make the payment.

The two simple models in Fig. 8.1, widely used in applications, have a single channel in which incoming vehicles receive the service and from which they depart when the service is completed. The order in which a vehicle is served (performing the maneuver or making the payment) coincides with the order of arrival, i.e. First In-First Out (FIFO). As we will say in this chapter, different situations can occur: models with a higher number of channels in which the service takes place, or even different rules of service can be defined.

We specify that in highway engineering the intersection in Fig. 8.1a is of great interest. In fact, all methods for calculating the performance of unsignalized intersections are derived from the simple waiting system in Fig. 8.1a.

8.1 Queuing Systems: Variables and Basic Relationships

In this and in the following sections we will refer to the following characteristic variables of the waiting systems (Figs. 8.1b and 8.3):

- q(t): rate of vehicle arrivals upstream of the queue at the time t;
- $T_s(t)$: service time of the *i*th vehicle;
- c(t): capacity, at the time t, of the serving facility (known as a server). Capacity is equal to reciprocal value of the service time, i.e. $c(t) = 1/T_s(t)$;
- $\rho(t) = q(t)/c(t)$: degree of saturation, also called traffic intensity;
- A(t): number of vehicles that have entered the system up to the time t (cumulative value of incoming vehicles, i.e. arrivals);

- D(t): number of vehicles that have left the system up to the time t (cumulative value of vehicles served, i.e. departures).

In a generic waiting system, arrivals and departures can occur over time on a regular basis (deterministic processes) or randomly (random processes). In the latter case, arrivals and departures may have different laws of probability (Chap. 5). These probabilities can be variable or constant over time. In general, we assume that arrivals occur completely randomly (Poissonian). As for the process of departures (the succession of users served), for service time $T_S(t)$ we usually consider exponential distributions with rate c(t), or generic distributions with known mean $E[T_S]$ and variance $VAR[T_S]$.

These variables allow us to study the waiting system, using the following state variables that evolve over time:

- L_C: queue length (number of vehicles queuing);
- L_S: number of vehicles in the system (number of vehicles queuing plus the one in service);
- w_C: time spent in the queue (between the vehicle arrival and the start of its service);
- w_S: time spent in the system (between the vehicle arrival and its departure upon completion of the service).

Given the physical meaning of w_S, w_C and T_S, it results: $w_S = w_C + T_S$.

L_C, L_S, w_C and w_S are random variables and the evolution over time of each of them forms the respective random process. Applications typically refer to the mean values of L_C, L_S, w_C and w_S, that are $E[L_C]$, $E[L_S]$, $E[w_C]$ and $E[w_S]$.

$E[L_S]$ and $E[w_S]$ are usually preferred, because they contain more information than $E[L_C]$ and $E[w_C]$, and the other variables can be derived in the various cases from them.

A useful point of view for the study of queues is provided by the analysis of the properties of the individual realization (sample-path) of the waiting system, i.e. a generic realization of A(t) and D(t) during an observation time interval T [3]. Once the cumulative variables A(t) and D(t) had realization—arrivals and departures had happened randomly during T—they are known without uncertainty. In this way, in a retrospective glance, the states of the system are determined in all instants of T. This is the case, for example, of a traffic video recording lasting T, which allows us to determine the succession over time of the number of arrivals and departures for queue systems in Fig. 8.1. Thus, for the complete knowledge of the behavior of the system the probabilistic aspect becomes irrelevant.

A situation similar to the previous one occurs whenever the system evolves deterministically. This means that in a succession of observation periods, all of the same duration T, the sample-paths representing the accumulations of arrivals are all equal to each other. Similarly, the sample-paths that represent the cumulative departures in T are all equal to each other.

Typically, if the evolution of the wait system is deterministic, the cumulative variables A(t) and D(t) are predetermined sequences of step-functions as shown in

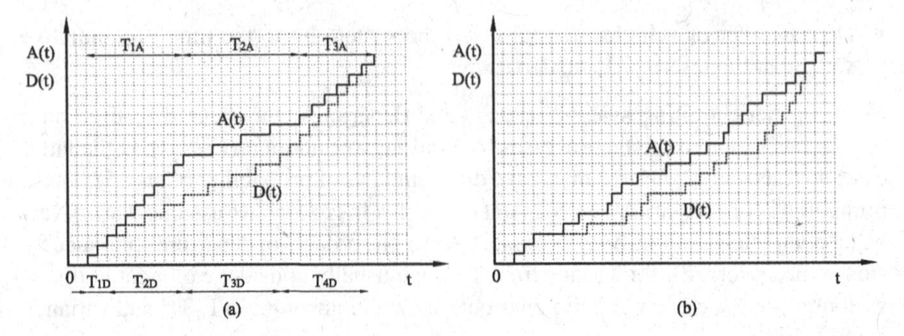

Fig. 8.2 Cumulated arrivals A(t) and departures D(t) step-function: **a** regular for intervals; **b** non-regular

Fig. 8.2 (e.g. regular for intervals). In step-functions, jumps, all equal to the unit, represent the arrivals or departures that occur over time.

Ultimately, in the deterministic case, for each instant t the cumulated number A(t) of vehicles that will enter the system and the cumulated number D(t) of vehicles that will exit the system up to t are known without uncertainty.

If an initial queue is not present, we obtain the number of vehicles queuing at the time t simply as a balance between the two cumulative functions:

$$L_s(t) = A(t) - D(t) \tag{8.1}$$

with $D(t) \leq A(t)$ for each time instant t.

Equation (8.1) is valid both if arrivals and departures are random and if they are deterministic.

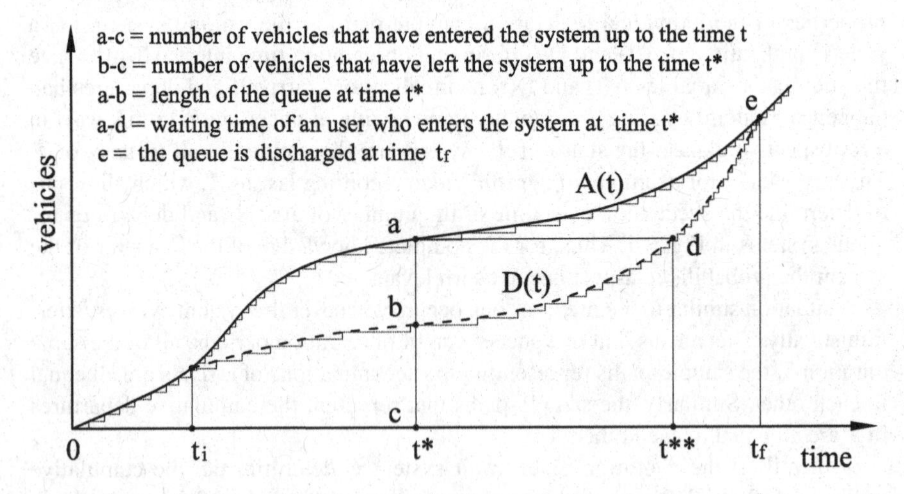

Fig. 8.3 Fluid approximation for A(t) and D(t)

For applications, A(t) and D(t) interpolate with continuous time functions (fluid approximation). In this case, the arrivals rates q(t) = dA(t)/dt and departure rates c(t) = dD(t)/dt also vary continuously and without random fluctuations. With fluid approximation, the study of waiting phenomena is greatly simplified.

In addition, with fluid approximation, both for a deterministic system and for a random system, we obtain some relationships between the state variables of the waiting system easily. This is because fluid approximation is valid for both deterministic and random waiting systems.

With reference to Fig. 8.3, we observe that at any given time t* the length of the queue is equal to the length of the ab segment, that is parallel to the ordinate axis between A(t) and D(t). Similarly, the waiting time of an user who enters the system at any given time t* is equal to the length of the segment ad, that is parallel to the abscissa axis from A(t*) to D(t**), i.e. the difference between the exit and the arrival instants of a given vehicle.

The total time W_s spent in the system by all vehicles during the time interval [0, t] graphically corresponds to the area between curves A(t) and D(t) in the same time range:

$$W_s(t) = \int_0^t N(t) \cdot dt = \int_0^t [A(t) - P(t)] \cdot dt = \int_0^t \left[q(t) \cdot t - c(t) \cdot t\right] \cdot dt \quad (8.2)$$

The average delay $\overline{w}_s$ for each user in the time interval [0, t] is the ratio of the total W_s delay to the number of vehicles arrived in the same interval:

$$\overline{w}_s = \frac{W_s(t)}{A(t)} = \frac{\int_0^t \left[q(t) - c(t)\right] \cdot t \cdot dt}{\int_0^t q(t) \cdot dt} \quad (8.3)$$

The average number of users in the system in the time interval [0, t] is therefore:

$$\overline{L}_s(t) = \frac{W_s(t)}{t} = \frac{\int_0^t \left[q(t) - c(t)\right] \cdot t \cdot dt}{\int_0^t c(t) \cdot dt} \quad (8.4)$$

8.1.1 Little's Law

Equation (8.4) can also be expressed in the following form:

$$L_s(t) = \frac{W_s(t)}{t} = \frac{W_s(t)}{A(t)} \cdot \frac{A(t)}{t} = \overline{w}_s \cdot \overline{q}(t) = \overline{w}_s \cdot q \qquad (8.5)$$

The previous expression is known as Little's Law. It describes a general behavior of the waiting system and links the average number of users in the system, the average waiting time and the average rate of arrivals. Referring to the queue, we can write similarly:

$$L_c(t) = \overline{w}_c \cdot q \qquad (8.6)$$

From Eq. (8.5) we obtain:

$$\overline{w}_s = \frac{L_s(t)}{q} \qquad (8.7)$$

8.2 Operating Conditions and Models for Waiting Systems

In general, queue theory models relate to different operating conditions for waiting systems.

Operating conditions are distinguished in stationary (or steady-state) conditions and in non-stationary (or transitory) conditions.

We assume that in a stationary state the arrival rate q(t) and the capacity of the system c(t) can vary randomly, while their average values q and c remain indefinitely constant over time.

Queue theory models for stationary conditions are applicable when the average rate of arrivals is less than the average capacity ($q < c$). In addition, q and c must be constant for a sufficiently long time from the beginning of the phenomenon. Under these conditions, the solutions that are obtained for state variables are called equilibrium (or steady-state) solutions and they are time-independent.

Stationary conditions are generally studied and applied for the analysis of communication or production systems. In fact, in these fields the basic conditions of the stationary state are largely achievable in most real cases.

In highway engineering, on the other hand, non-stationary situations are prevalent. In fact, the traffic demand and/or capacity of a transport system vary over time during the periods of the day.

In all cases where the system becomes more and more significantly congested ($\rho \gg 1$), arrivals and departures are less and less random. Thus, the waiting systems tend to behave more and more deterministically. In these conditions, fluid approximation solutions are used.

If a waiting system is not in a stationary state but is not even in a situation of marked congestion ($\rho > 1$), queue theory provides probabilistic solutions. These time-dependent probabilistic solutions, however, need too complicated mathematics for practical calculations.

Time-dependent heuristic solutions have been developed in the field of highway engineering. These heuristic solutions are easier to use than probabilistic solutions.

Heuristic solutions allow us to analyze waiting systems in a unified and continuous way. In fact, these solutions are valid from low-intensity situations of traffic ($\rho \ll 1$) up to congestion ($\rho \geq 1$).

The following sections are dedicated to: probabilistic models for stationary conditions; deterministic models and heuristic models for non-stationary conditions.

8.3 Probabilistic Models for Steady-State

The most commonly used queue theory solutions relate to stationary systems. These practical solutions are applicable if q(t) and c(t) vary randomly but with constant average values over time.

Strictly speaking, these solutions apply if the system is unsaturated ($\rho = q/c < 1$) and after an indefinitely long time since the beginning of the waiting phenomenon. This time is known as *queue relaxation time*. After this time, in fact, the expected values (i.e. the mean values) $E[L_c]$, $E[L_s]$, $E[w_c]$ and $E[w_s]$ are those of statistical equilibrium. So, beyond the relaxation time these values are constant. In this case, $E[L_c]$, $E[L_s]$, $E[w_c]$ and $E[w_s]$ depend only on the average values of q and c and on their probability laws.

We consider completely random arrivals (Poissonians) with constant rate q. So, the headways between the arriving vehicles are distributed exponentially (Chap. 5). Instead, more than one probability distribution can be assumed for service time T_s.

For practical purposes, as a useful approximation, the relaxation time value T_r is provided by Morse's relationship [4]. Morse's relaxation time is valid for completely random arrivals and exponential service times. According to Morse, the waiting system is in a stationary state if q and c (with $q/c < 1$) are constant in such a T-time that is:

$$T > T_r = 1/(\sqrt{c} - \sqrt{q})^2 \tag{8.8}$$

This equation is also used for different distributions of T_s [5, 6].

Therefore, in a stationary state, with service time T_s following a generic probability distribution, P-K formulas (Pollaczek-Khinchine [1, 2]) are used. In particular, for the cases presented in Fig. 8.1, the average waiting time in the system is:

$$E[w_s] = E[T_s] + \frac{q \cdot (E[T_s]^2 + VAR[T_s])}{2 \cdot (1 - q \cdot E[T_s])} \tag{8.9}$$

With Eqs. (8.5) and (8.9) we can determine the average number of users in the system $E[L_s]$ on the minor road or in the tollbooth lane of Fig. 8.1:

$$E[L_s] = q \cdot E[T_s] + \frac{q^2 \cdot (E[T_s]^2 + VAR[T_s])}{2 \cdot (1 - q \cdot E[T_s])} \tag{8.10}$$

If service times are distributed according to an Erlang probability distribution with parameter K, $E[T_s]$ and $VAR[T_s]$ are:

$$E[T_s] = T_c + \frac{e^{K \cdot q_1 \cdot T_c} - \sum_{i=0}^{K} \frac{(K \cdot q_1 \cdot T_c)^i}{i!}}{q_1 \cdot \sum_{i=0}^{K} \frac{(K \cdot q_1 \cdot T_c)^i}{i!}} \tag{8.11}$$

$$VAR[T_s] = \frac{(K+1) \cdot \left[e^{K \cdot q_1 \cdot T_c} - \sum_{i=0}^{K+1} \frac{(K \cdot q_1 \cdot T_c)^i}{i!} \right]}{K \cdot q_1^2 \cdot \sum_{i=0}^{K-1} \frac{(K \cdot q_1 \cdot T_c)^i}{i!}} + (E[T_s] - T_c)^2 \tag{8.12}$$

with T_c critical interval for the maneuver performed by the minor stream.

For the intersection in Fig. 8.1, we can use the values of Table 8.1 for K in Eqs. (8.11) and (8.12), depending on the flow rate Q on the major road.

Equations (8.11), (8.12) may be particularized in the case of exponential or deterministic (constant) service time T_s.

For exponential service time, we have $VAR[T_s] = (E[T_s])^2$.

For deterministic service time (i.e. constant) we have $VAR[T_s] = 0$.

With these values for the variances of T_s, by Eqs. (8.9) and (8.10) with $c = 1/E[T_s]$ and $\rho = q/c$ we obtain the expressions in Table 8.2.

We get $E[L_c]$ and $E[w_c]$ from the relationship of immediate meaning $E[w_s] = E[w_c] + E[T_s] = E[w_c] + 1/c$ and by Eqs. (8.5) and (8.6). So we get:

$$E[w_c] = E[w_s] - 1/c \tag{8.13}$$

$$E[L_c] = E[L_s] - \rho \tag{8.14}$$

Table 8.1 Erlang distribution parameter K depending on Q on the major road

Q (veh/h)	K value
0 ÷ 500	1
501 ÷ 1.000	2
1.001 ÷ 1.500	3

Table 8.2 Average values of waiting times and number of vehicles in the system (stationary state)

Probability distribution		System state variables	
Arrivals	Service time	$E[w_s]$	$E[L_s]$
Poissonian	Exponential	$\frac{1}{c\cdot(1-\rho)}$	$\frac{\rho}{(1-\rho)}$
Poissonian	Constant	$\frac{2-\rho}{2\cdot c\cdot(1-\rho)}$	$\frac{2\rho-\rho^2}{2\cdot(1-\rho)}$

Finally, with Eqs. (8.13) and (8.14) we get the expressions of Table 8.3.

As mentioned above, these cases relate to the situations in Fig. 8.1, characterized by a single serving channel. In Table 8.4 we also report the average values of the state variables for a waiting system with random arrivals (Poissonian) and exponential service times, in the presence of m serving channels [1, 2]. In this case we consider $\rho = q/(m \cdot c)$.

It should be noted that the expressions in Table 8.4 apply in case of a single waiting line for the service, which in turn is carried out in parallel on m different channels. However, this configuration is not common in highway engineering. In fact, in a freeway toll plaza there are more parallel service points (tollbooths) and multiple parallel queues (lanes). These situations cannot be dealt with Table 8.4 solutions. It is necessary to use more articulated solutions, often attributable to multiple and mutually interacting single-channel queueing systems. A simple case in which, however, the equations in Table 8.4 can be used is a freeway toll plaza with m tollbooths of

Table 8.3 Average values of waiting times and number of vehicles in the queue (stationary state)

Probability distribution		Queue state variables	
Arrivals	Service time	$E[w_c]$	$E[L_c]$
Poissonian	Exponential	$\frac{\rho}{c\cdot(1-\rho)}$	$\frac{\rho^2}{(1-\rho)}$
Poissonian	Constant	$\frac{\rho}{2\cdot c\cdot(1-\rho)}$	$\frac{\rho^2}{2\cdot(1-\rho)}$

Table 8.4 State variables for waiting system with m serving channels, for random arrivals and exponential service times

Variable	Poissonian arrivals and exponential service times for m serving channels
Prob. of zero vehicles in the system (n = 0)	$P_0 = \left[\sum_{K=1}^{m-1} \frac{(m\cdot\rho)^K}{K!} + \frac{(m\cdot\rho)^K}{m!} \cdot \left(\frac{1}{1-\rho}\right)\right]^{-1}$
Prob. of n vehicles in the system	$P_n = P_0\cdot(\rho\cdot m)^n/n!$ if $n = 1, 2, m-1$ $P_n = P_0\cdot(\rho\cdot m)^n/(m^{n-m}\ m!)$ if $n = m, m+1, \ldots$
Ave. Num. veh. in the queue	$E[L_c] = P_0\cdot(\rho\cdot m)^{m+1}/[m!\ m(1-\rho)^2]$
Ave. Num. veh. in the system	$E[L_s] = E[L_c] + (\rho \cdot m)$
Ave. Waiting time in the system	$E[w_s] = E[L_s]/q$
Ave. Waiting time in the queue	$E[w_c] = E[L_s]/q - 1/c$

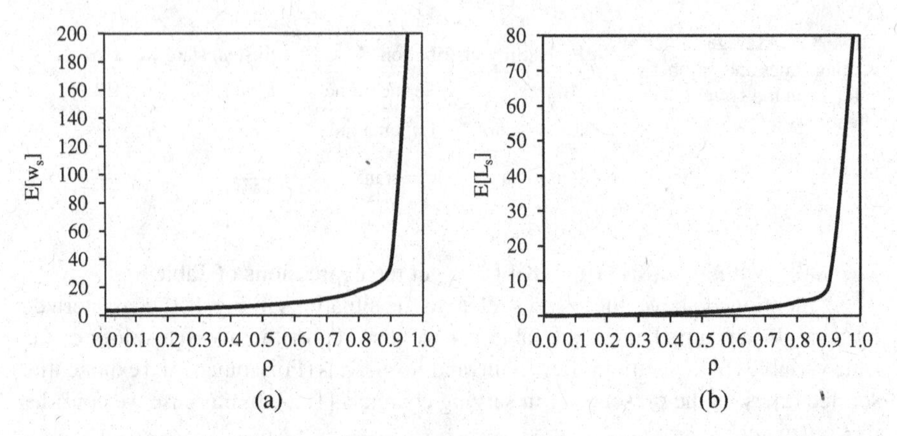

Fig. 8.4 Poissonian queue with exponential time service ($c = 1000$ veh/h)—state variables with varing ρ: **a** $E[w_s]$ in seconds; **b** $E[L_s]$

uniform features. In this case, in fact, users waiting for service in the m lanes can be considered as being part of a single queue.

Going back to the general case, we can deduce directly from the observation of real systems when a waiting system can be considered stationary. This happens if q and c can be considered constants in a sufficiently long time interval [cf. Eq. (8.8)]. In these conditions, the formulas in Tables 8.2 and 8.3 are valid for $E[L_c]$, $E[L_s]$, $E[w_c]$ and $E[w_s]$.

Anyway, based on these formulas $E[L_c]$, $E[L_s]$, $E[w_c]$ and $E[w_s]$ tend quickly to ∞ (cf. Fig. 8.4) as ρ grows, typically starting at $\rho = 0.8$. This result is clearly unrealistic. In fact, the time intervals for which we have $\rho \geq 0.8$, although long, are always of finite duration, so the queue cannot grow indefinitely. Therefore, in place of Tables 8.2 and 8.3 solutions, approximate solutions (i.e. heavy traffic solutions [2] or heuristics solutions [5]) are used.

Example 1 A freeway toll plaza has only one tollbooth where only one payment method is available. From traffic surveys it has been estimated that the average flow of users arriving at the toll plaza is $q = 410$ veh/h and that the average service time for users' payment is $E[T_s] = 8$ s. Arrivals and service times are distributed according to a Poisson law and an exponential law, respectively. We want to calculate the main performance indicators of the system.

The capacity of the system is:

$$c = 1/E[T_s] = 1/8 = 0.125 \text{ veh/s} = 450 \text{ veh/h}$$

The degree of saturation is:

$$\rho = q/c = 410/450 = 0.91 < 1 \text{ (system under sub} - \text{saturation conditions)}$$

The average queue length is:

$$E[L_c] = \rho^2/(1-\rho) = 0.91^2/(1-0.91) = 9.3 \text{ veh}$$

The average number of vehicles in the system is:

$$E[L_s] = \rho/(1-\rho) = 0.91/(1-0.91) = 10.2 \text{ veh}$$

The average waiting time in the queue is:

$$E[w_s] = \rho/[c \cdot (1-\rho)] = 0.91/[360 \cdot (1-0.91)] \cdot 3600 = 82 \text{ s.}$$

The average waiting time in the system is:

$$E[w_s] = 1/[c \cdot (1-\rho)] = 1/[360 \cdot (1-0.91)] \cdot 3600 = 90 \text{ s}$$

From the point of view of the performance of the toll facility, we feel that this number of queueing vehicles is too high. For this reason, we want to assess the effects of the introduction of an additional tollbooth next to the existing one. We hypothesize to introduce a new tollbooth with the same functionality as the existing one. Even for this tollbooth, therefore, service times have exponential distribution with mean $E[T_s] = 8$ s.

The waiting system we want to evaluate will still have random arrivals (Poissonian) and exponential service times, in the presence of $m = 2$ serving points. As the capacity is 450 veh/h for each of the two tollbooths, the saturation level of the overall system is:

$$\rho = q/(m \cdot c) = 410/(2 \cdot 450) = 0.46$$

Applying the formulas in Table 8.4, the probability that there are no vehicles in the system (with $m = 2$) is:

$$P_0 = (1-\rho)/(1+\rho) = (1-0.46)/(1+0.46) = 0.374$$

The average number of vehicles in the system is:

$$E[L_s] = 2\rho/(1-\rho^2) = 2 \cdot 0.46/\left(1-0.46^2\right) = 1.15 \text{ veh}$$

The average waiting time in the system is:

$$E[w_s] = 1/\left[c \cdot (1-\rho^2)\right] = 1/\left[450 \cdot (1-0.46^2)\right] \cdot 3600 = 10.09 \text{ s}$$

The average queue length is:

$$E[L_c] = 2\rho^3/(1-\rho^2) = 2 \cdot 0.46^3/\left(1-0.46^2\right) \cdot 3600 = 0.24 \text{ veh}$$

The average waiting time in the queue is:

$$E[w_s] = \rho^2/[c \cdot (1 - \rho^2)] = 0.46^2/[450 \cdot (1 - 0.46^2)] \cdot 3600 = 2.09 \text{ s}$$

The opening of a new tollbooth to manage traffic demand on the freeway toll plaza allows to halve the degree of saturation. The average number of users in the system is drastically reduced (from 10.2 to 1.15 vehicles), as well as the average waiting time in the system (from 90 to 2 s per vehicle).

8.4 Deterministic Solutions in Congestion

When traffic demand is much greater than the capacity value ($q \gg c$), waiting systems in Fig. 8.1 are over-saturated ($\rho \gg 1$).

For these conditions, the waiting line does not reach the steady state. The systems are not in statistical equilibrium and therefore $E[L_c]$, $E[L_s]$, $E[w_c]$ and $E[w_s]$ vary over time.

In conditions of growing over-saturation, both arrivals and departures are less and less random. Systems tend to behave, therefore, in a deterministic manner. In other words, in over-saturated conditions, ($\rho > 1$), the more ρ grows the more the mean values of the state variables are close to the corresponding deterministic values.

State variables can be described with a time-dependent deterministic model. We obtain the expressions for the number of vehicles in the system, the total and the average waiting time considering the expressions (8.4) (8.2) and (8.3). Considering the presence of a nonzero initial number of waiting vehicles in the system, that is $L_s(t = 0) = L_{s0} \neq 0$ assuming that q and c can be considered constant in the [0, t] interval and the total and average waiting times calculated up to the instant t, we have:

$$L_s(t) = L_{s0} + (q - c) \cdot t = L_{s0} + (\rho - 1) \cdot c \cdot t \tag{8.15}$$

$$W_s(t) = L_{s0} \cdot t + (q - c) \cdot \frac{t^2}{2} = L_{s0} \cdot t + (\rho - 1) \cdot c \cdot \frac{t^2}{2} \tag{8.16}$$

$$\overline{w}_s(t) = \frac{W(t)}{A(t)} = \frac{L_{s0}}{q} + \frac{(q - c)}{q} \cdot \frac{t}{2} = \frac{L_{s0}}{q} + (1 - \frac{1}{\rho}) \cdot \frac{t}{2} \tag{8.17}$$

If the initial number of waiting vehicles is zero, the number of vehicles in the system, the total delay and the average queue are obtained by simply placing $L_{s0} = 0$ in Eqs. (8.15), (8.16) and (8.17).

Expressions (8.4) (8.2) and (8.3) were obtained by considering a time interval [0, t] where q and c are constants. More general situations with q and c not constants on a time interval [0, T] can still be treated with the same expressions. In fact, we can break down the whole [0, T] interval into a series of sub-intervals $[t_{i-1}, t_i]$ in which q_i and c_i can be considered constants [6] (see Example 4).

The Eqs. (8.15) and (8.17) clearly show a non-stationary state of the waiting system. In fact, the number of vehicles and the average waiting time increase linearly as time progresses. In this sense, deterministic solutions (8.15) and (8.17) are time-dependent solutions.

A classic example of this type of evolutionary state are traffic bottlenecks. Traffic bottlenecks originate in highway sections, due to the presence of work zones or accidents. In a certain section of the highway, there is a change in capacity from the value $c > q$ to the value $c_1 < q{\cdot}t^*$ is the duration of the interval in which we see the reduction of capacity from c to c_1 $(t = t_0 + t^*)$, while t_f is the instant for which we can observe the end of the effects induced by the reduction of capacity (final queue discharge from instant t).

Figure 8.5 shows a diagram for identifying the relationships between the different variables under consideration. This type of diagram is always very useful in solving deterministic queue problems. Table 8.5 shows some useful expressions for describing waiting phenomena at bottlenecks, considering an initial state with $L_s(0) = 0$.

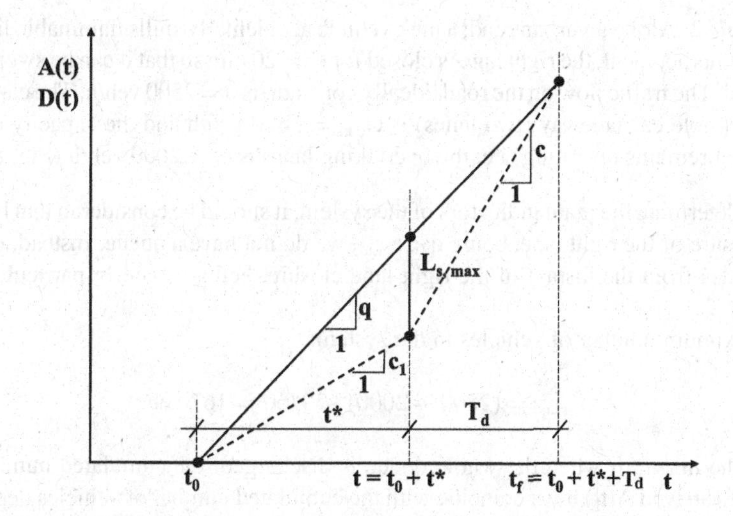

Fig. 8.5 Deterministic queue diagram for a traffic bottleneck

Table 8.5 System indicators for capacity changes in a traffic bottleneck

Variable	Expression
Max number of vehicles in the system	$L_{s,max} = (q - c_1){\cdot}t^*$
Queue discharge instant	$t_f = (c - c_1) \cdot t^*/(c - q)$
Total delay	$(q - c_1) \cdot t^* \cdot t_f/2$
Total queued vehicles	$q \cdot t_f$

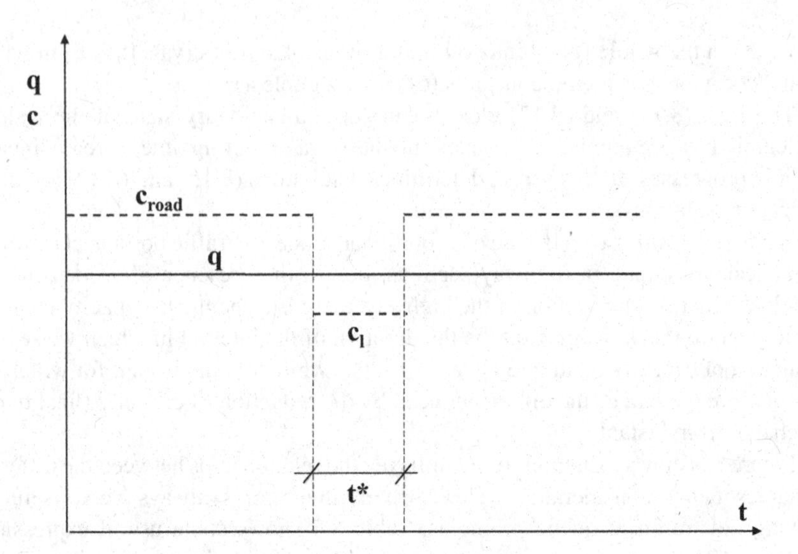

Fig. 8.6 Time evolution of traffic demand and capacity

Example 2 Along an urban road, a tank vehicle accidentally spills flammable liquid. Due to this accident, the right lane is closed for $t^* = 20$ min so that it can be swept and cleaned. The traffic flow on the road, ideally constant, is $q = 2500$ veh/h. The capacity of the whole carriageway (two lanes) is $c_{road} = 3800$ veh/h and the capacity of the lane that remains operating (i.e. the overtaking lane) is $c_l = 2000$ veh/h (Fig. 8.6).

To determine the main indicators of the system, it should be considered that before the closure of the right lane, being $q < c_{road}$, we do not have a queue. Instead, queue originates from the instant of the right lane closure, being $q > c_l$. In particular, we have:

Maximum number of vehicles in the system:

$$L_{s,max} = (2500 - 2000) \cdot 20/60 = 167 \text{ veh}$$

In the instant t_f when the whole queue is discharged, the cumulated number of vehicles arrived $A(t_f)$ must coincide with the cumulated number of vehicles departed $D(t_f)$. We have:

$$A(t_f) = q \cdot t_f$$
$$D(t_f) = c_l t^* + c_{road} \cdot (t_f - t^*)$$

It must be $A(t_f) = D(t_f)$, and then we get:

$$t_f = (c_{road} - c_l)/(c_{road} - q) \cdot t^* = (3800 - 2000)/(3800 - 2500) \cdot 20 = 28' = 1680 \text{ s}$$

The total users' delay is:

$$W = (q - c_l)\frac{t^* \cdot t_f}{2} = (2500 - 2000) \cdot \frac{(20/60) \cdot (28/60)}{2} = 38.9 \text{ h}$$

The average user's delay corresponds to the area between A(t) and D(t) for the time interval $[t, t_f]$; after a few calculations we get:

$$\overline{w}_s = \frac{(q - c_l)}{q} \cdot \frac{T^*}{2} = \frac{2500 - 2000}{2500} \cdot \frac{20}{2} = 2'$$

The number of queued vehicles is:

$$N = q \cdot t_f = 2500/3600 \cdot 1680 = 1166 \text{ veh}$$

The average number of vehicles in the system is:

$$L_s = W/t_f = 38.9/(1680/3600) = 83.36 \text{ veh}$$

Example 3 Near a roundabout with four arms, each with two entry lanes and one exit lane, a shopping center will be built. We expect that this new shopping center will produce an increased traffic demand. For the future project configuration, we estimate a traffic demand outgoing from one of the arms of 800 veh/h during the hourly interval 7:00–8:00 A.M. and 1600 veh/h during the 8:00–9:00 A.M. hourly interval, while the exit capacity of the same arm is 1200 veh/h.

Considering the portion of the roundabout ring lane (which has a single lane with a length of L = 120 m) between this arm and the previous one, we want to estimate how long it takes to become saturated during the time interval 8:00–9:00 A.M.

To answer this question, we assume an average vehicle length of 4.5 m and a safety gap between pairs of vehicles of 0.5 m. In these hypotheses, the portion of the roundabout ring lane is able to accommodate a queue of $N = 120 / (4.5 + 0.5) = 24$ veh.

Since in the time interval 7:00–8:00 A.M. we have $q > c$, at the beginning of the second hourly interval there are no vehicles in the queue. But from 8:01 A.M., it occurs that $q > c$, so a queue of N vehicles would arise in a time $t^* = N/(q - c) = 24/(1600 - 1200) = 0.06 \text{ h} = 216 \text{ s}$.

This queue congest the ring lane and the whole intersection.

8.5 Heuristic Solutions for Steady and Non-steady States

We have already pointed out that, for the study of queue systems, in stationary and under-saturated conditions ($\rho < 1$) probabilistic models are applied. On the other hand, in non-stationary over-saturated conditions ($\rho \gg 1$) we can use deterministic models as useful approximations.

When the degree of saturation is $\rho > 1$, but the flow isn't much bigger than the capacity (that is, it doesn't happen that $\rho \gg 1$) the deterministic approach is not entirely suitable to be employed. In these cases, the random effects of arrivals and departures are not entirely negligible.

On the other hand, in under-saturated conditions ($\rho < 1$), stochastic models do not allow us to study several non-stationary problems that arise in reality. This is the case, for example, in which the flow rates vary over time and exceed the value of capacity for a certain period, but always with $\rho < 0.8$.

In technical practice, therefore, in stationary conditions but with $\rho > 0.6 - 0.8$ and in non-stationary conditions, heuristic solutions are used. These solutions calculate the state variables of the waiting systems in Fig. 8.1, replacing the complex non-stationary probabilistic formulations.

For a given state parameter δ of the queue system (e.g. δ can represent the average number of vehicles or the average waiting time in the system) heuristic solutions are continuous functions of the type $\delta^* = f(\rho)$. These functions are continuous function of ρ, which can take on all values in the range $[0, +\infty]$. In this way, heuristic solutions allow us to analyze in a unified way the systems in Fig. 8.1 from low-traffic intensity situations ($\rho \ll 1$) to congestion conditions ($\rho \geq 1$).

Heuristic solutions $\delta^* = f(\rho)$ are identified by combining the expected value $E[\delta]$ of a state variable in a stationary state with its deterministic value $\overline{\delta}(\rho)$.$\overline{\delta}(\rho)$ is time-dependent, so the whole heuristic solution $\delta^*(\rho)$ is time-dependent. $\delta^*(\rho)$ is obtained by imposing that $f(\rho)$ has as an oblique asymptote $\overline{\delta}(t)$ [6]. For this reason, the functions $\delta^* = f(\rho)$ that we get for the different state variables are also called queue transition functions.

From an application point of view, heuristic solutions can be identified by the coordinate transformation method [7]. To exemplify, let's consider the case of the average number of users in the system. Trends for the average number of users in the system $E[L_s]$ and $\overline{L}_s(t)$ are showed in Fig. 8.7. By applying the coordinate transformation, we impose the condition:

$$\rho_d - \rho_t = 1 - \rho_s \tag{8.18}$$

Here we deal with the case of random arrivals (Poissonian) and exponential service times. Similar transformations can also be made for different service times. $\overline{L}_s(t)$ is given by Eq. (8.15):

$$\overline{L}_s(t) = L_{s,0} + (\rho_d - 1) \cdot c \cdot t \Rightarrow \rho_d = \frac{\overline{L}_s(t) - L_{s,0}}{c \cdot t} + 1 \tag{8.19}$$

The probabilistic expression for $E[L_s]$ is:

$$E[L_s] = \frac{\rho_s}{(1 - \rho_s)} \Rightarrow \rho_s = \frac{E[L_s]}{1 + E[L_s]} \tag{8.20}$$

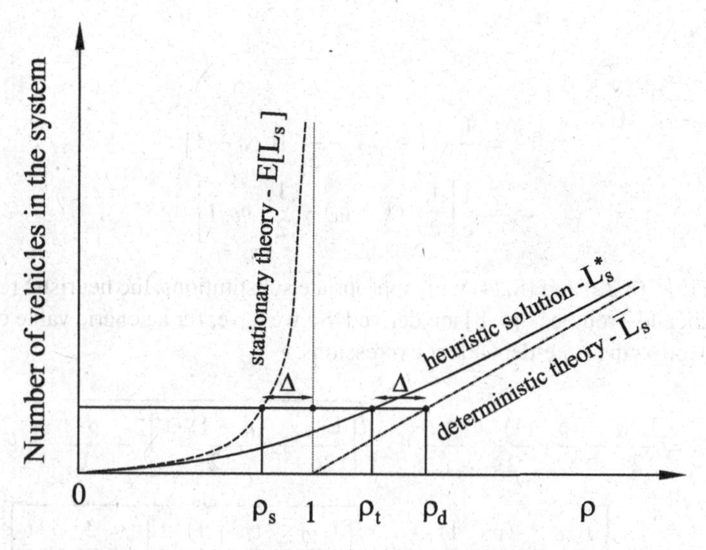

Fig. 8.7 Transition curve for the average number of users in the system L_s^*

With Eqs. (8.19) and (8.20), and considering that for ρ_t, ρ_d, ρ_s we have $L_s^* = \overline{L}_s = E[L_s]$, it results that:

$$\frac{L_s^* - L_{s,0}}{c \cdot t} + 1 - \rho_t - 1 + \frac{L_s^*}{1 + L_s^*} = 0 \tag{8.21}$$

The Eq. (8.21) is equivalent to the following expression:

$$L_s^{*2} + b' \cdot L_s^* - c' \tag{8.22}$$

where

$$b' = 1 + c \cdot t - c \cdot t \cdot \rho_t - L_{s,0};$$
$$c' = c \cdot t \cdot \rho_t + L_{s,0}$$

The solution for Eq. (8.22) is:

$$L_s^* = \frac{1}{2}\left[-b' \pm \sqrt{b'^2 + 4 \cdot c'}\right] \tag{8.23}$$

In the same way, considering $E[w_s]$ and $\overline{w}_s$, we get the heuristic solution for $w_s{}^*$:

$$w_s^* = \frac{1}{2}\left[-b'' \pm \sqrt{b''^2 + 4 \cdot c''}\right] \tag{8.24}$$

where:

$$b'' = \frac{t}{2} \cdot (1 - \rho_t) - \frac{1}{c} \cdot \left[L_{s,0} + 1\right]$$
$$c'' = \frac{1}{c}\left[\frac{t}{2} \cdot (1 - \rho_t) + \frac{1}{2} \cdot \rho_t \cdot t\right]$$

From Eqs. (8.22) and (8.24) with appropriate substitutions, the heuristic relations of Akçelik and Troutbeck [5, 8] are derived. So we have, for a generic value of ρ_t ($\rho_t = \rho$) the following time-dependent expressions:

$$L_s^* = \frac{L_{s,0}}{2} + \frac{(\rho - 1) \cdot c \cdot t}{2} + c \cdot \sqrt{\left[\frac{L_{s,0}}{2c} + \frac{(\rho - 1) \cdot t}{2}\right]^2 + \frac{\rho \cdot t}{c}} \tag{8.25}$$

$$w_s^* = \frac{1}{c} + \frac{1}{2} \cdot \left[\frac{L_{s,0}}{c} + \frac{(\rho - 1) \cdot t}{2} + \sqrt{\left[\frac{L_{s,0}}{c} + \frac{(\rho - 1) \cdot t}{2}\right]^2 + \frac{2 \cdot \rho \cdot t}{c}}\right] \tag{8.26}$$

If we consider $L_{s,0} = 0$ in Eq. (8.26), we get the heuristic solution for the average waiting time in the system suggested by HCM [9, 10] for unsignalized intersections:

$$w_s^* = \frac{1}{c} + \frac{t}{4} \cdot \left[(\rho - 1) + \sqrt{(\rho - 1)^2 + \frac{8 \cdot \rho}{c \cdot t}}\right] \tag{8.27}$$

All preceding formulations are valid if traffic variables (q and c) are expressed in vehicles (veh/s or veh/h). If, as often occurs, q and c are expressed in pcu (pcu/s or pcu/h) specific corrections must be introduced in order to have consistent results [11]. If q' is the flow expressed in veh/h and q is the flow expressed in pcu/h (considering the appropriate values of the equivalence coefficients for the interested vehicular classes), considering $f = q/q'$, Eqs. (8.25) and (8.26) are modified as follows [8]:

$$L_s^* = \frac{L_{s,0}}{2} + \frac{(\rho - 1) \cdot c \cdot t}{2} + c \cdot \sqrt{\left[\frac{L_{s,0}}{2c} + \frac{(\rho - 1) \cdot t}{2}\right]^2 + \frac{f \cdot \rho \cdot t}{c}} \tag{8.28}$$

$$w_s^* = \frac{f}{c} + \frac{1}{2} \cdot \left[\frac{L_{s,0}}{c} + \frac{(\rho - 1) \cdot t}{2} + \sqrt{\left[\frac{L_{s,0}}{c} + \frac{(\rho - 1) \cdot t}{2}\right]^2 + \frac{2 \cdot f \cdot \rho \cdot t}{c}}\right] \tag{8.29}$$

The Eq. (8.27), on the other hand, can be rewritten as follows [11]:

$$w_s^* = \frac{1}{c} + \frac{t}{4} \cdot \left[(\rho - 1) + \sqrt{(\rho - 1)^2 + \frac{8 \cdot f \cdot \rho}{c \cdot t}}\right] \tag{8.30}$$

To get q expressed in pcu per hour, equivalence coefficients (passenger car equivalent, PCE) are generally used in international technical practice (see Chap. 1). In the case of roundabout entries, for example, HCM [9] suggests PCE values equal to 1.0 for Passenger Car (PC) and 2.0 for Heavy Vehicle (HV).

Example 4 Let's consider a freeway toll plaza with a single tollbooth used for credit card payments. The service time for the toll payment follows an exponential probability distribution with mean $E[T_s] = 8$ s. The average service capacity is then $c = 1/E[T_s] = 3600/8 = 450$ veh/h.

We want to study the evolution of the system in a 2.5 h time interval T. During T traffic demand varies in sub-intervals T_1, T_2, T_3, T_4 as in Fig. 8.8.

In $T_1 = 1$ h we have $q_1 = 300$ veh/h. The intensity of traffic in T_1 is $\rho_1 = 300/450 = 0.667$. The relaxation time of the queue [Eq. (8.8)] is $T_r = 1/(\sqrt{450/3600} - \sqrt{300/3600})^2 = 238$ s. Being $\rho < 1$(and widely less than 0.8) and $T_1 > T_r$, waiting system can be considered in a stationary state. Applying the relationships in Table 8.2 we obtain, at the end of T_1, $E[L_s{}^{(1)}] = 2$ veh and $E[w_s{}^{(1)}] = 24$ s.

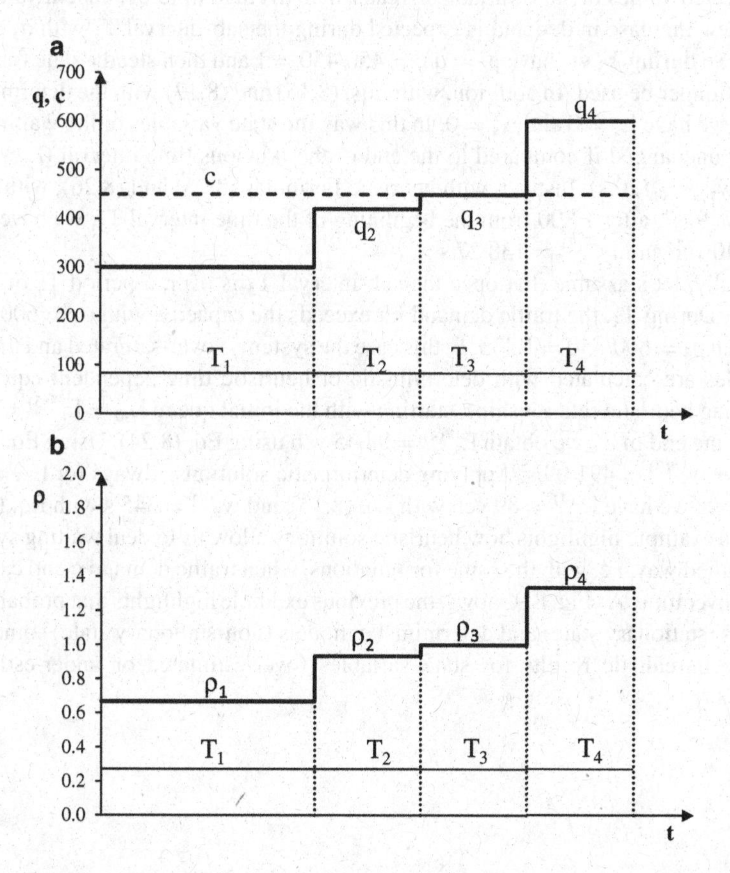

Fig. 8.8 Time evolution of the freeway tollbooth waiting system

Practically coincident results can be achieved by applying the heuristic solutions (8.25) and (8.26), considering no waiting vehicles at the beginning of T_1. After 3,600 s from the beginning of the phenomenon, in fact, we will have considering the Eq. 8.25 $L_s^{*(1)} = 1.97$ veh, and $w_s^{*(1)} = 23.59$ s considering the Eq. (8.26).

After T_1, in the next half hour T_2 traffic demand grows up to $q_2 = 420$ veh/h, with $\rho_2 = 420/450 = 0.933 < 1$. In this case, with the formulas in Table 8.2, at the end of T_2 we have $E[L_s^{(2)}] = 14$ veh and $E[w_s^{(2)}] = 120$ s. On the other hand, with heuristic solutions (8.25) and (8.26), with $L_s^{*(1)}$ equal to the vehicles waiting at the beginning of T_2 ($L_s^{*(1)} = 2$), at the end of T_2 (1800 s), we get $L_s^{*(2)} = 9.37$ veh and $w_s^{*(2)} = 70.80$ s.

The values obtained with heuristic solutions are different from those obtained with the steady-state formulas. As we have already noted in this chapter, with $\rho_2 > 0.8$ (as in this case) expressions in Tables 8.2 and 8.3 give unrealistic results. In addition, with the specified values for q_2 and c during T_2, the relaxation time for the steady-state queue given by the Eq. (8.8) is just under two hours (6958 s). This value is well above the duration of T_2 (1800 s). Thus, stationary probabilistic formulations result in expected values of state variables greater than average time-dependent values.

A new increase in demand is expected during the sub-interval T_3 with $q_3 = 450$ veh/h. So during T_3 we have $\rho_3 = q/c = 450/450 = 1$ and then steady-state formulas can no longer be used. In addition, with Eqs. (8.15) and (8.17) with the deterministic model we have $L_s = 0$ and $w_s = 0$. In this way the state variables of the waiting line appear unchanged if compared to the end of the previous time interval ($L_s^{(3)} = 14$ veh e $w_s^{(3)} = 120$ s). Instead, with heuristic Formulas (8.25) and (8.26), with $L_{s0} = L_s^{*(2)} = 9.37$, after 1,800 from the beginning of the time interval T_3, we have $L_s^{*(3)} = 20.40$ veh and $w_s^{*(3)} = 138.27$ s.

Finally, let's assume that once the sub-interval T_3 is over, a period T_4 of 1800s begins. During T_4, the traffic demand far exceeds the capacity, with $q_4 = 600$ veh/h and then $\rho_4 = 600/450 = 1.333$. In this case the system is over-saturated and the state variables are calculated with deterministic or heuristic time-dependent equations. Applying heuristic relationships, starting with the initial queue $L_{s0} = L_s^{*(3)} = 20.40$ veh, at the end of T_4 we obtain $L_s^{*(4)} = 98.45$ veh using Eq. (8.24). Using Eq. (8.26) we have $w_s^{*(3)} = 491.09$ s. Applying deterministic solutions, always for $L_{s0} = L_s^{(3)} = 14$ veh, we have $L_s^{(4)} = 89$ veh with Eq. (8.15) and $w_s^{(4)} = 345$ s with Eq. (8.17).

This example highlights how heuristic solutions allow us to deal waiting systems in a unified way, i.e. with the same formulations, when traffic demand q and capacity c vary over time. As Fig. 8.9 shows, the previous example highlights that probabilistic models (stationary state) and deterministic models (non-stationary state) sometimes lead to unrealistic results for state variables (over-estimated or under-estimated values).

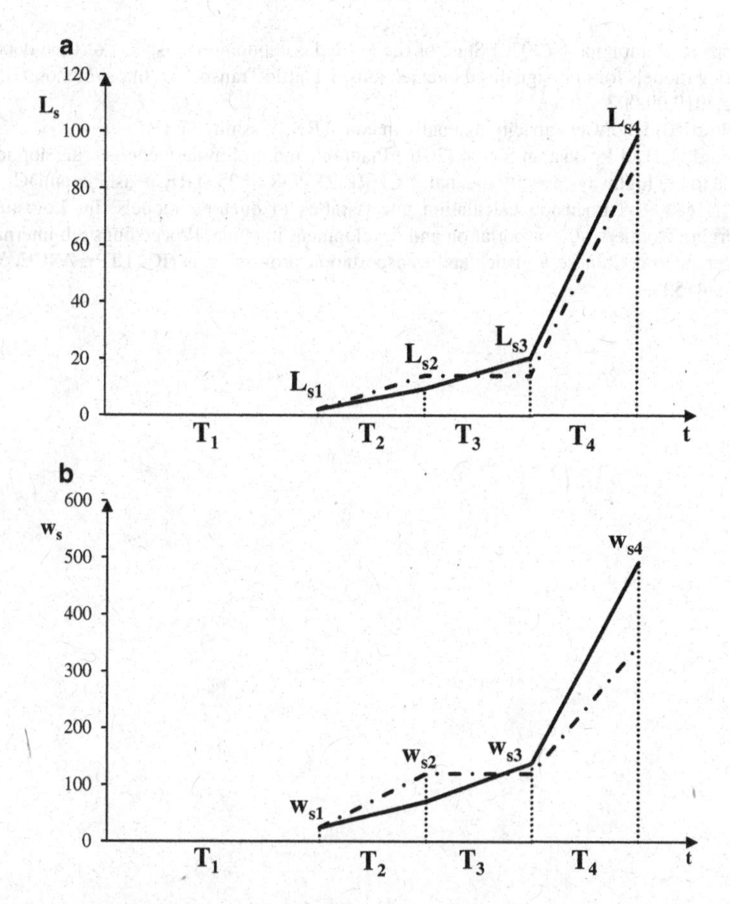

Fig. 8.9 (**a**) Number of vehicles in the system L_S and (**b**) average waiting time w_S at the end of each interval (continuous line, heuristic solutions; dotted line, probabilistic solutions for stationary state and deterministic solutions for non-stationary state)

References

1. Kleinrock L (1976) Queueing systems. In: Volume 2: computer applications. Wiley, New York
2. Shortle JF, Thompson JM, GrossD, Harris CM (2018) Fundamentals of queueing theory, vol. 399. Wiley & Sons
3. El-Taha M, Stidham S Jr (1992) Deterministic analysis of queueing systems with heterogeneous servers. Theoret Comput Sci 106(2):243–264
4. Morse PM (1962) Queues, inventories and maintenance. Wiley
5. Troutbeck RJ, Brilon W (2000) Unsignalized intersection theory. In: Gartner N, Messner CJ, Rathi AK (eds) Traffic flow theory. TRB Special Report, TRB, Washington, DC
6. Mauro R (2010) Calculation of roundabouts. Springer
7. Kimber RM, Hollis EM (1979) Traffic queues and delays at road junctions. Transport and Road Research Laboratory Report No. LR909. TRRL, Crowthorne, UK

8. Mauro R, Pompigna A (2020) State of the art and computational aspects of time-dependent waiting models for non-signalised intersections. J Traffic Transp Eng. https://doi.org/10.1016/j.jtte.2019.09.007
9. TRB (2010) Highway capacity manual, 5th edn. TRB, Washington DC
10. Richard D, Paul R, Bastian S et al (2016) Planning and preliminary engineering applications guide to the highway capacity manual. NCHRP REPORT 825. TRB, Washington DC
11. Wu N (2009) Dimension calculation uncertainties in queuing models. In: Logistics: the emerging frontiers of transportation and development in china, Proceedings 8th international conference on Chinese logistics and transportation professionals (ICCLTP), ASCE, VA, pp 4745–4753

Chapter 9
Unsignalized Intersections

Abstract This chapter deals with unsignalized intersections and methods for determining the measures of effectiveness (MOE): waiting times and delays. To this regard, the TRL method for estimating capacity, queues, delays at three-arm intersections is exemplified.

The measures of effectiveness (MOE) of an unsignalized at-grade intersection (controlled by *yield* or *stop signs*) are determined by estimating delays and queues.

In order to calculate queues it needs hypothesizing the probability laws for service time and vehicle arrivals.

In an intersection there can be identified a minor street and a major street and, consequently, a minor stream and a major stream.

Should flows be modest on the minor street, the arrival process generally follows the Poisson law (headways according to an exponential variable). On the other hand, should flows be high, the use of Erlang distribution appears to be more frequent.

For service times (Sect. 9.1) the following considerations can be made:

- in *unsignalized intersections* the possibility for the minor street drivers to cross the traffic stream depends on the distribution of time headways of the stream. This distribution is correlated to the service time;
- in *signalized intersections* the entries from minor to major streets depend on the green and red times which are constant in time (with the only exception of the Self-Adaptive Traffic Signal systems, capable of regulating the signal timing parameters in real time according to the fluctuation of traffic demand). Therefore, for signalized intersections with conventional traffic signals, the service time is a deterministic variable.

That said, two different situations can occur, as already described in Chap. 8:

- if there is a gap between the arrival flow in the system and the capacity ($q < c$), *the queue is stationary,* that is, it randomly varies over time but within a limited value range (in other words, it cannot grow indefinitely);

M. Guerrieri and R. Mauro, *A Concise Introduction to Traffic Engineering*,
Springer Tracts in Civil Engineering,
https://doi.org/10.1007/978-3-030-60723-4_9

- if the arrival flow in the system is equal or higher than the capacity ($q \geq c$), *the queue is not stationary* and its length extends over time growing indefinitely.

9.1 Waiting Times and Delays in Unsignalized Intersections

In a generic intersection, vehicle trajectory (time-distance diagram, see Figs. 1.1 and 1.2, Chap. 1) appears as schematized in Fig. 9.1.

Since the flow is interrupted (see Sect. 1.1), for each vehicle the following basic traveling and delay phases can be identified [1]:

- *traveling phase with the regime speed* approaching the intersection (corresponding to e.g. design speed), with no waste of time;
- *deceleration time*: necessary to pass from the regime speed to speed equal to zero;
- *waiting (queuing) time* w_c: it is estimated from the time instant when the vehicle reaches the queue to the instant when the vehicle is set on the yield (or stop) line;
- *service time* T_s: corresponding to the waiting time at the queue head. In other terms, it is the time taken between the instant when the vehicle reaches the yield (or stop) line and the instant when it starts performing the entry or crossing maneuver at the intersection;
- *acceleration time*: time during which the vehicle accelerates to reach the regime speed;
- *motion phase with the regime speed* going far away from the intersection (corresponding to, e.g. design speed).

Thus, it is possible to define, as already done in Chap. 8, the following traffic parameters:

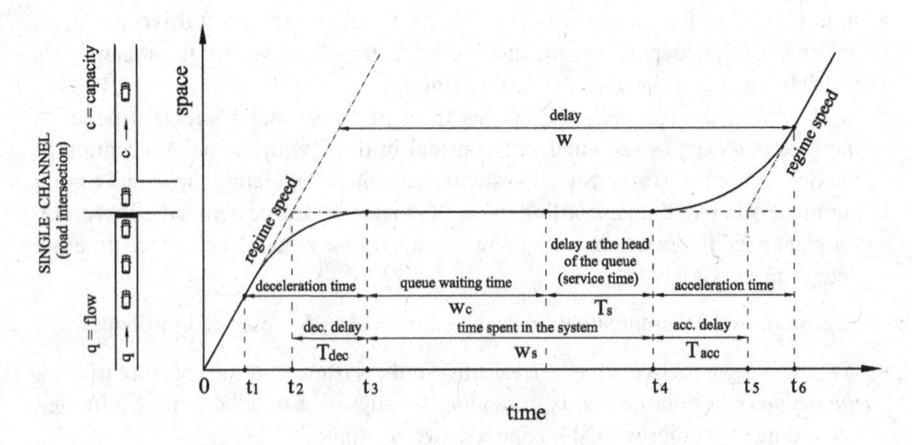

Fig. 9.1 Vehicle trajectory and delay due to the presence of an intersection

- *Waiting time in the system* w_s: equal to the sum of the queuing time (w_c) and service time (T_s):

$$w_s = w_c + T_s \tag{9.1}$$

- *Control delay at the intersection* w (also known as total delay): equal to the sum of the *waiting time in the system* (w_s) the deceleration delay T_{dec} and acceleration delay T_{acc} (see Fig. 9.1):

$$w = w_s + T_{dec} + T_{acc} \tag{9.2}$$

If $E[w_s]$ denotes the average waiting time in the system, taking the relation (9.1) into account, it yields:

$$E[w_s] = E[w_c] + E[T_s] = E[w_c] + \frac{1}{c} \tag{9.3}$$

In a steady-state condition (see Chap. 1), the mean (average) $E[\cdot]$ is a mathematical expectation of a random variable.

Moreover, Little's formula (see Chap. 8) makes it possible to determine the average number of queuing users $E[L_s]$:

$$E[L_s] = q \cdot E[w_s] \tag{9.4}$$

9.2 The Levels of Service for Unsignalized At-Grade Intersections

The level of service (LOS) of at-grade intersections, including roundabouts, is determined in function of the control delay w (computed or measured) and is defined for each minor movement.

Similarly to what examined in Chap. 2, also for intersections the level of services are six (A to F). More in detail, the Highway Capacity Manual (HCM 2016) defines the thresholds shown in Table 9.1.

Table 9.1 Level-of-Service criteria for unsignalized at-grade intersections

Average control delay w [s/veh]	Level of service (LOS)	
	Flow-to-capacity ratio $\rho = q/c \leq 1$	Flow-to-capacity ratio $\rho = q/c > 1$
0–10	A	F
> 10–15	B	F
> 15–25	C	F
> 25–35	D	F
> 35–50	E	F
> 50	F	F

9.3 Capacity, Delay and Queue at Three-Arm Intersections

The methodology here suggested is the method proposed by *Transport Research Laboratory* (TRL). Its advantages consist in the simplicity for calculation together with the possibility of monitoring, at the intersection, the queue length evolution over time, in accordance with the heuristc solutions for time-dependent models (see Chap. 8). Unlike the methods based on the *gap acceptance theory,* indicated in the HCM, the TRL method does not require employing users' critical gap and follow-up time which are difficult to be determined experimentally (see Chap. 7).

The TRL method can be applied if the operating speed (v_{85}) on the main road does not exceed 85 km/h. The input data of the TRL method are traffic flows and the intersection geometry.

The capacities of maneuvers b–a, b–c and c–b in Fig. 9.2 can be calculated by expressions [2–4]:

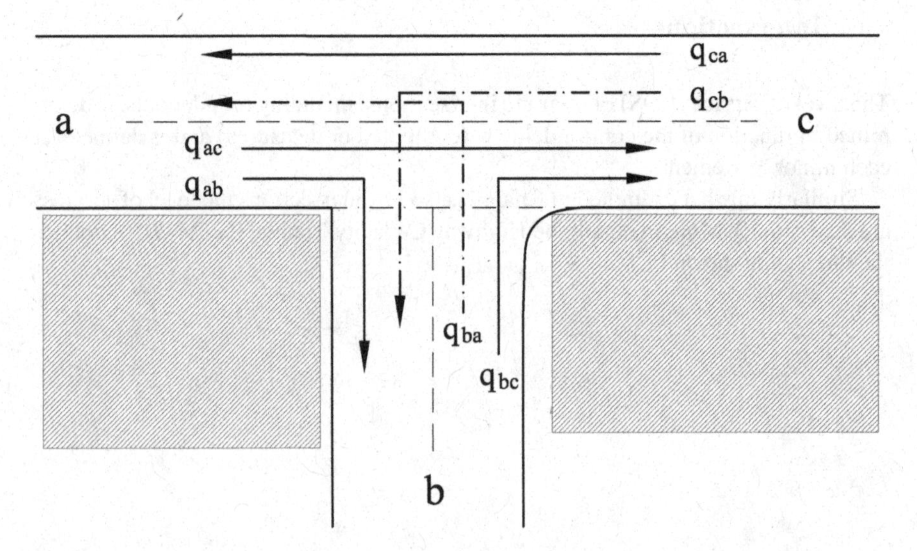

Fig. 9.2 Traffic streams in a T-intersection (a–c Main Street)

$$c_{ba} = D \cdot \left[627 + 14 \cdot W_{cr} - Y\left(0.364 \cdot q_{ac} + 0.114 \cdot q_{ab} + 0.229 \cdot q_{ca} + 0.520 \cdot q_{cb}\right)\right] \quad (9.5)$$

$$c_{bc} = E \cdot \left[745 - Y \cdot \left(0.364\, q_{ac} + 0.114 \cdot q_{ab}\right)\right] \quad (9.6)$$

$$c_{cb} = F \cdot \left[745 - 0.364 \cdot Y\left(q_{ac} + q_{ab}\right)\right] \quad (9.7)$$

The coefficients D, E and F of expressions (9.5), (9.6), (9.7) are calculated by the following relations:

$$D = [1 + 0.094(w_{ba} - 3.65)] \cdot [1 + 0.0009(Vs_{ba} - 120)] \cdot [1 + 0.0006(Vd_{ba} - 150)] \quad (9.8)$$

$$E = [1 + 0.094(w_{bc} - 3.65)] \cdot [1 + 0.0009(Vs_{bc} - 120)] \quad (9.9)$$

$$F = [1 + 0.094(w_{cb} - 3.65)] \cdot [1 + 0.0009(Vs_{cb} - 120)] \quad (9.10)$$

Since:

- c_{ba}, c_{bc}, c_{cb} are the capacities (in pcu/h) of turning maneuvers in Fig. 9.2;
- q_{ac}, q_{ab}, q_{ca}, q_{cb} are the measured or estimated flows (in pcu/h);
- w_{ba}, w_{bc}, w_{cb} are the average width of the lanes on which the maneuvers b-a, b-c and c-b are respectively performed. They are determined on a 20 m segment starting from the yield or stop line.

If there are two lanes on the minor street, assume w_{ba} as the average width of the right-turning lane and w_{bc} as the average width of the left-turning lane.

If, on the other hand, there is a single shared lane of average width $w = (a + b + c + d + e)/5$ (see Fig. 9.3), it follows: $w_{ba} = w_{bc} = w/2$.

Similar consideration can be made for w_{cb}, in that there may be a single shared lane for manouvres c–b and c–a or two dedicated lanes on the major street;

- Vs_{ba}, Vs_{bc}, Vs_{cb} stand for the available sight distance on the left-hand side of the users carrying out the maneuvers b–a, b–c and c–b;
- Vd_{ba} denotes the available sight distance on the right-hand side of drivers carrying out the manouver b–a. The available sight distance is determined by considering that the road user's eye height is 1.10 m and that the user is set at 10 m from the yield or stop line;
- W_{cr} is the width of the traffic island on the major street (if present);
- $Y = (1\text{–}0.0345 \cdot W)$, with W denoting the total width of the carriageway of the major street.

The geometrical dimensions are all expressed in metres.

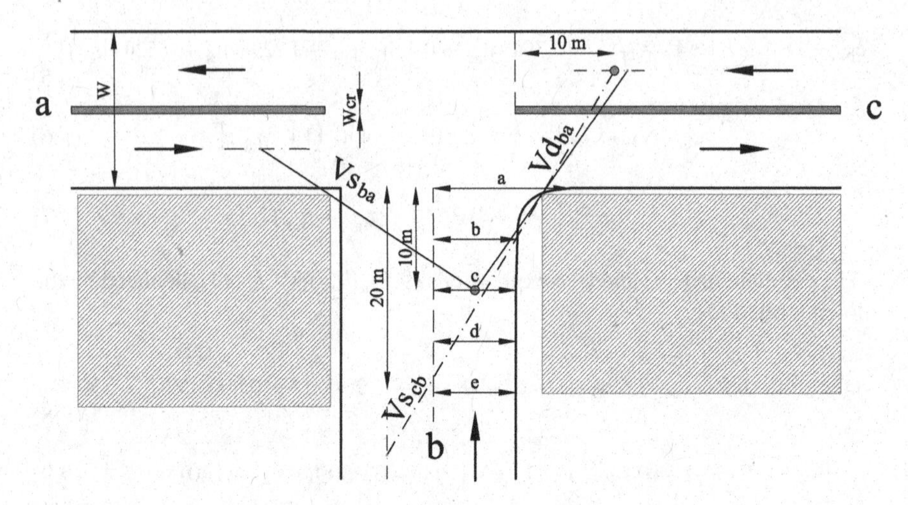

Fig. 9.3 Characteristic dimensions and visibility distances to be considered

After calculating the capacities with expressions (9.5), (9.6) and (9.7), the saturation degree ($\rho_{i,j} = q_{i,j}/c_{i,j}$) is determined for each maneuver "ij". Therefore, expressions (8.23) and (8.27) enable the calculation of the queue length and the average waiting time of every vehicle respectively.

9.3.1 Case Study: Delay Calculation in a T-Intersection

Determine delays and queues in a T-intersection by considering an analysis time t = 1 h.

The intersection has the following geometry (see Fig. 9.2) and traffic demand:

- minor street with a single entry lane onto the major street of constant width near the yield line: a = b = c = d = and = 5.00 m;
- major street with total carriageway width: W = 9.50 m and without traffic island;
- $Vs_{ba} = Vs_{bc} = Vd_{ba} = 30$ m, $Vs_{cb} = 50$ m;
- $q_{ca} = 790$ pcu/h; $q_{ac} = 560$ pcu/h; $q_{ab} = 170$ pcu/h; $q_{ba} = 200$ pcu/h; $q_{bc} = 145$ pcu/h; $q_{cb} = 110$ pcu/h.

The average lane width of the minor street is:

$$w = (5.00 + 5.00 + 5.00 + 5.00 + 5.00)/5 = 5.00\text{m}$$

Since the lane is shared, it follows:

$$w_{ba} = w_{bc} = w/2 = 5.00/2 = 2.50\text{m}$$

The half-carriageway width of the major street is 9.50/2 = 4.75 m.

It has a single lane (flows q_{ca} and q_{cb}); therefore it follows: $w_{cb} = 4.75/2 = 2.375$ m.

Moreover it results:

$$Y = 1 - 0.0345 \cdot 9.50 = 0.672$$

With these data, through expressions (9.8), (9.9) and (9.10), it yields:

$$D = 0.77, E = 0.82, F = 0.82$$

Expressions (9.5), (9.6) and (9.7) enable the calculation of the capacities c_{ba}, c_{bc}, c_{cb} and, consequently, the saturation degrees ($\rho_{ba} = q_{ba}/c_{ba}$; $\rho_{bc} = q_{bc}/c_{bc}$; $\rho_{cb} = q_{cb}/c_{cb}$).

Finally, after determining the coefficients b', c', b'' and c'', expressions (8.23) and (8.27) make it possible to obtain the queue length L_s* and the average delay of every vehicle w_s* respectively.

The synthetic results are shown in the table below under the hypothesis that at the beginning of the analysis period there is no queuing vehicle ($L_{s,0} = 0$).

For the manouver b-a there are 4 vehicles in the queue. Moreover, since the average vehicle delay is equal to 70 s/veh, the level of service of this stream is F (see Table 9.1).

Maneuver	q (pcu/h)	c (pcu/h)	ρ	b'	c'	L_s* (veh)	b''	c''	w_s* (s/veh)
c-a	790	-	-	-	-	-	-	-	-
a-c	560	-	-	-	-	-	-	-	-
a-b	170	-	-	-	-	-	-	-	-
b-a	200	244	0.82	45.14	200.00	4.06	0.09	0.002	69.76
b-c	145	488	0.30	343.65	145.00	0.42	0.35	0.001	10.48
c-b	110	467	0.24	358.09	110.00	0.31	0.38	0.001	10.06

9.4 Capacity, Delays and Queues at Roundabout Intersections

9.4.1 *Entry Capacities*

In conventional roundabouts (Fig. 9.4) [5] the entry capacity depends first and foremost on the number of the circulating lanes and entries, and on user behaviors.

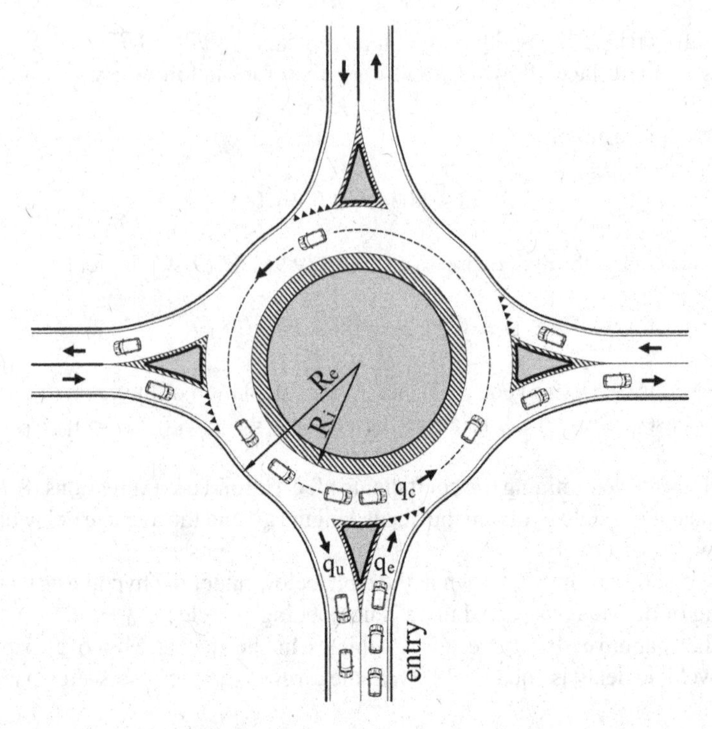

Fig. 9.4 Entry flow (q_e), exit flow (q_u) and circulating flow (q_c)

Below the well-known Brilon-Wu formula is illustrated (Eq. 9.11) [6]; it is derived from the gap-acceptance theory (see Chap. 7). It is also applied in the German guidelines on roundabouts [7]:

$$c = 3600 \cdot \left(1 - \frac{T_{min} \cdot q_c}{n_k \cdot 3600}\right)^{n_k} \cdot \frac{n_z}{T_f} \cdot e^{-\frac{q_c}{3600}\left(T_c - \frac{T_f}{2} - T_{min}\right)} \tag{9.11}$$

where

- c = entry capacity of the analysed arm (pcu/h);
- q_c = circulating flow in front of the entry in question (pcu/h);
- n_k = number of circulating lanes;
- n_z = number of entry lanes;
- T_c = critical gap;
- T_f = follow-up time;
- T_{min} = minimum headway between succeeding vehicles on the circulating carriageway.

Generally, the following values are assumed as users' psycho-technical parameters: $T_c = 4.1$ s; $T_f = 2.9$ s; $T_{min} = 2.1$ s. By means of these values and expression

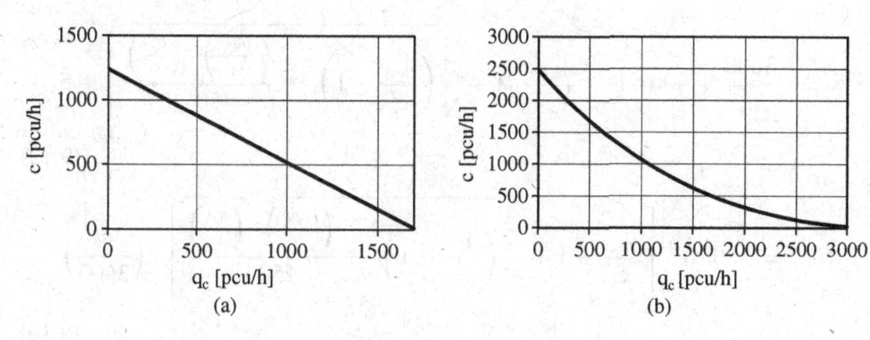

Fig. 9.5 Entry capacity in function of the circulating flow: **a** roundabouts with 1 entry lane and 1 circulating lane; **b** roundabouts with 2 entry lanes and 2 circulating lanes

Table 9.2 Capacity values of the circulating carriageway [1]

Roundabout type	Number of entry lanes	Circulating carriageway capacity [pcu/h]
Roundabouts with 1 circulating lane (mini roundabouts and compact roundabouts)	1	1600
Compact roundabouts with 2 circulating lanes	1	1600
	2	1600
Large roundabouts	1	2000
	2	2500

(9.11), the c capacity values in function of the circulating flow q_c can be obtained as represented in Fig. 9.5 for roundabouts with one entry lane and one circulating lane, and for roundabouts with two lanes at entry and two lanes on the circulating carriageway.

The characteristic capacity values of the circulating carriageway are indicated in Table 9.2 [1].

On the other hand, the capacity of 1-lane exits is of the order of 1200 pcu/h [1].

9.4.2 *Control Delays and Queues*

Through the Origin/Destination (O/D) matrices of the measured or estimated traffic demand (see Chap. 2) it is easy to calculate the entry ($q_{e,i}$), exit ($q_{u,i}$) and circulating ($q_{c,i}$) flows for the i-th arm of the roundabout. Relation (9.11) makes it possible to determine the entry capacity (c_i) and the pertaining saturation degree ($\rho_i = q_{e,i}/c_i$).

These data also enable the calculation of the average control delay (w_i) and the queue lengths ($N_{95,i}$) by applying the following expressions [8]:

$$w_i = \frac{3600}{c_i} + 900 \cdot T \cdot \left[\frac{q_{e,i}}{c_i} - 1 + \sqrt{\left(\frac{q_{e,i}}{c_i} - 1 \right)^2 + \frac{\left(\frac{3600}{c_i} \right) \cdot \left(\frac{q_{e,i}}{c_i} \right)}{450 \cdot T}} \right] + 5 \tag{9.12}$$

$$N_{95,i} = 900 \cdot T \cdot \left[\frac{q_{e,i}}{c_i} - 1 + \sqrt{\left(\frac{q_{e,i}}{c_i} - 1 \right)^2 + \frac{\left(\frac{3600}{c_i} \right) \cdot \left(\frac{q_{e,i}}{c_i} \right)}{150 \cdot T}} \right] \cdot \left(\frac{c_i}{3600} \right) \tag{9.13}$$

where

- w_i denotes the average vehicle delay of the i-th arm (s/veh);
- $N_{95,i}$ denotes the 95th percentile of the queue length (veh) in the reference interval T;
- $q_{e,i}$ stands for the traffic flow from the arm i (veh/h);
- c_i denotes the entry capacity (veh/h);
- T is the time period, h (T = 1 for a 1-h analysis, T = 0.25 for a 15-min analysis);
- 5 is the time constant per vehicle (s/veh) which considers the accelerations and decelerations due to the queue and priority to circulating vehicles.

Once known the average control delay at each entry, the level of service of every entry is determined with Table 9.1.

The roundabout average delay (w_r) is calculated as a weighted average of the entry delays (w_i), by assuming the flows $q_{e,i}$ as weights:

$$w_r = \frac{\sum_i w_i \cdot q_{e,i}}{\sum_i q_{e,i}} \tag{9.14}$$

Once known w_r, the average level of service for the roundabout can be obtained by using Table 9.1 again.

9.4.3 Effect of the Pedestrian Flow on the Entry Capacity

In urban roundabouts, the pedestrian flows lead to capacity reductions at entries and exits that are proportional to the pedestrian traffic intensity.

In the presence of pedestrian crosswalks, the entry capacity can be estimated mostly by using the three calculation procedures [1]: English (Marlow and Maycock), French (CETE) and German (Brilon, Stuwe and Drews).

These calculation procedures can be applied only if the pedestrian flow (crossing the roundabout arm orthogonally) has the priority over the vehicle traffic flow.

Fig. 9.6 Examples of turbo-roundabouts

According to the German procedure, the entry capacity with pedestrian crossings c_{ped} can be obtained by multiplying the capacity in the absence of pedestrian crossings c (calculated, for instance, with Eq. (9.11)) and a reduction factor M (< 1); the reduction factor M takes the pedestrian flow into account:

$$c_{ped} = c \cdot M \tag{9.15}$$

M is calculated in function of the entry layout:

- 1-lane entry:

$$M = \frac{1119.5 - 0.715 \cdot q_c - 0.644 \cdot q_{ped} + 0.00073 \cdot q_c \cdot q_{ped}}{1069 - 0.65 \cdot q_c} \tag{9.16}$$

- 2-lane entry:

$$M = \frac{1260.6 - 0.381 \cdot q_{ped} - 0.329 \cdot q_c}{1380 - 0.50 \cdot q_c} \tag{9.17}$$

where

- q_c denotes the circulating flow on the circulating carriageway opposite the entry in question (pcu/h);
- q_{ped} is the pedestrian traffic flow, crossing from the arm in question (ped/h).

9.4.4 Non Conventional Roundabouts: Turbo-Roundabouts

The most well-known non-conventional roundabouts are spiral roundabouts, those with by-pass lanes for right-turning (and/or for crossing intersections), flower roundabouts and turbo-roundabouts (Fig. 9.6) [9].

Turbo-roundabouts are more and more used in several European countries (e.g. Holland, Slovenia, etc.) and are distinguished by the typical “turbine” configuration of the central island (and the circulating carriageway), and by the physical separation of the lanes (with curbs).

Indeed, this layout provides potentially higher safety conditions than traditional roundabouts [10] and, in some traffic demand cases, also higher capacity values..

Some models for calculating capacity, delays and queues in non-conventional roundabouts are described in [11].

9.4.5 *Case Study: Determination of Levels of Service in a Large-Sized Roundabout*

The large-diameter roundabout in Fig. 9.7, (external diameter D = 100 m) has two circulating lanes. Entries1 and 2 have two lanes, while Entries 3, 4 and 5 have a single lane. Given the matrix O/D of the traffic demand in Fig. 9.7, the aim is to estimate the delays and levels of service for every entry and the average delay for the roundabout.

In order to solve the problem, first it needs determining the values of the flows entering and exiting from each entry and those circulating opposite each entry.

Every O/D matrix cell in Fig. 9.7 shows the flow $q_{i,j}$ from origin i" (entry) to the destination "j" (exit). In the case in question, there are 5 arms, and consequently i = 1, 2, 3, 4, 5 and j = 1, 2, 3, 4, 5.

For instance, the flow from arm 4 (origin) towards arm 3 (destination), is $q_{4,3}$ = 25 veh/h.

Denoted as such, the total entry flow from the i-th arm onto the roundabout can be obtained by summing up the values of all the cells in row "i". Similarly, in order to determine the flow exiting from the i-th arm, the values of all the cells in column "j" are to be summed up.

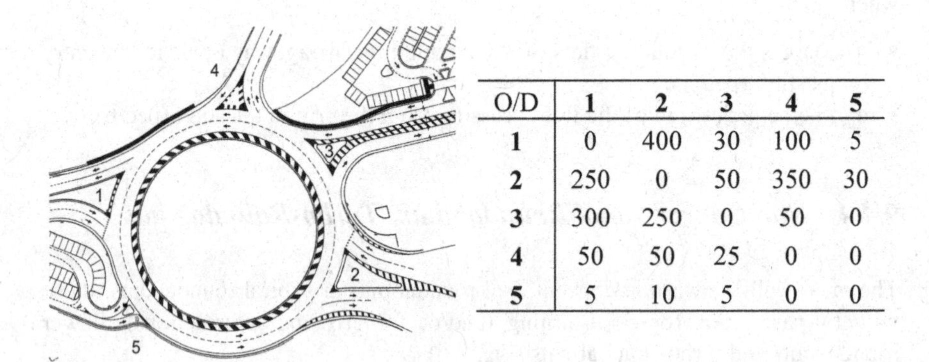

O/D	1	2	3	4	5
1	0	400	30	100	5
2	250	0	50	350	30
3	300	250	0	50	0
4	50	50	25	0	0
5	5	10	5	0	0

Fig. 9.7 Roundabout layout and origin/destination matrix (flows in veh/h)

For example, there is.

- Entry flow from arm 1:

$$q_{e,1} = q_{1,1} + q_{1,2} + q_{1,3} + q_{1,4} + q_{1,5} = 0 + 400 + 30 + 100 + 5 = 535 \text{ veh/h}$$

- Exit flow from arm 1:

$$q_{u,1} = q_{1,1} + q_{2,1} + q_{3,1} + q_{4,1} + q_{5,1} = 0 + 250 + 300 + 50 + 5 = 605 \text{ veh/h}$$

On the other hand, the total circulating flow opposite each entry can be obtained with the following expressions:

$$q_{c,1} = (q_{4,5} + q_{4,2} + q_{4,3}) + (q_{3,5} + q_{3,2}) + q_{2,5}$$
$$q_{c,5} = (q_{1,2} + q_{1,3} + q_{1,4}) + (q_{4,2} + q_{4,3}) + q_{3,2}$$
$$q_{c,2} = (q_{5,3} + q_{5,4} + q_{5,1}) + (q_{1,3} + q_{1,4}) + q_{4,3}$$
$$q_{c,3} = (q_{2,4} + q_{2,1} + q_{2,5}) + (q_{5,4} + q_{5,1}) + q_{1,4}$$
$$q_{c,4} = (q_{3,1} + q_{3,5} + q_{3,2}) + (q_{2,1} + q_{2,5}) + q_{5,1}$$

Once known the circulating flow values, the capacity for every entry can be calculated with Expression (9.11), by assigning, given the roundabout geometry (see Fig. 9.7), the following values: $n_k = 2$; $n_z = 2$ for entries 1 and 2; $n_z = 1$ for entries 3, 4 and 5; $T_c = 4.1$ s; $T_f = 2.9$ s; $T_{min} = 2.1$ s.

For instance, for Entry 1, being $q_{c,1} = 0 + 50 + 25 + 0 + 250 + 30 = 355$ veh/h, the entry capacity is as follows:

$$c_1 = 3600 \cdot \left(1 - \frac{2.1 \cdot 355}{2 \cdot 3600}\right)^2 \cdot \frac{2}{2.9} \cdot e^{-\frac{355}{3600}\left(4.1 - \frac{2.9}{2} - 2.1\right)} = 1890 \text{veh/h}$$

The saturation degree for Entry 1 is $\rho = 535/1890 = 0.28$. The average vehicle delay of the entry, calculated with (9.12) is $w = 8$ s/veh, which corresponds to level of service (LOS) A (see Table 9.1).

The roundabout average delay calculated with (9.14) is $w_r = 16$ s/veh which corresponds to level of service (LOS) C (see Table 9.1).

The total results, for all arms and for the roundabout in its entirety, are shown in Table 9.3.

Table 9.3 Results from the functionality analysis

Roundabout layout and performance indicators	Arm n°				
	1	2	3	4	5
N° entry lanes	2	2	1	1	1
N° exit lanes	2	2	1	2	1
Entry flows $q_{e,i}$ (veh/h)	535	680	600	125	20
Exit flows $q_{u,i}$ (veh/h)	605	710	110	500	35
Circulating flows $q_{c,i}$ (veh/h)	355	165	735	835	855
Capacity c_i (veh/h)	1890	2194	685	625	614
Saturation degree ρ_i	0.28	0.31	0.88	0.20	0.03
Entry delays w_i (s/veh)	8	7	36	12	11
Level-of-service of Entry	A	A	E	B	B
Average delay w_r (s/veh)	16				
Roundabout level-of-service	C				

References

1. Mauro R (2010) Calculation of roundabouts. Springer
2. Kimber RM, Coombe RD (1980) The traffic capacity of major/minor priority junction. TRRL Supplementary Report 582
3. Salter RJ (1989) Highway traffic analysis and design, Palgrave Macmillan UK
4. Rogers M, Enright B (2017) Highway engineering. Wiley, 3th Edition
5. Italian guidelines for the design of road intersections (D.M. 19/4/2006)
6. Brilon W, Wu N (2008) Kapazitaet von Kreisverkehren—Aktualisierung, (Capacity of roundabouts—actual solution). Strassenverkehrstechnik, N. 5:280–288
7. Forschungsgesellschaft für Straßen und Verkehrswesen Arbeitsgruppe Straßenentwurf, Merkblatt für die Anlage von Kreisverkehren (2006)
8. Highway Capacity Manual: HCM (2016) Transportation Research Board, Washington, D.C.
9. Tollazzi T (2015) Alternative Types of Roundabouts. Springer
10. Mauro R, Cattani M, Guerrieri M (2015) Evaluation of the safety performance of turbo-roundabouts by means of a potential accident rate model. Baltic J Road Bridge Eng 10(1):28–38
11. Tollazzi T, Mauro R, Guerrieri M, Renčelj M (2015) Comparative analysis of four new alternative types of roundabouts: "Turbo", "flower", "target" and "four-flyover" roundabout. Periodica Polytechnica Civil Eng 60(1):51–60

Chapter 10
Signalized Intersections

Abstract This chapter covers signalized intersections. The HCM model for the calculation and functional analysis of such intersections is illustrated in detail.

Traffic streams at these intersections are regulated with traffic signal control in order to:

- reduce delay to vehicles and pedestrians crossing through the intersection;
- impose certain selected traffic management policies;
- reduce crash frequency and/or severity.

The installation of a traffic light system is justified, in functional terms, for medium–high traffic demand values. Figure 10.1 [1] shows a criterion for choosing the intersection control type in function of two-way peak-hour volumes on the major and minor streets leading to the junction.

10.1 The Cycle Length

The cycle length (C) is the total time expressed in seconds required for an orderly complete sequence (for all approaches) of green (G), yellow (Y) and red (R) phases to be completed. During a traffic light cycle all vehicle streams at the intersection are served. The cycle time is given by the relation [2]:

$$C = G + I + R \tag{10.1}$$

since:

- G (s) is *the green time* during which the traffic lights (within a cycle C), for a certain movement or combination of movements (lane group), are set to green, thus permitting to go ahead;

M. Guerrieri and R. Mauro, *A Concise Introduction to Traffic Engineering*,
Springer Tracts in Civil Engineering,
https://doi.org/10.1007/978-3-030-60723-4_10

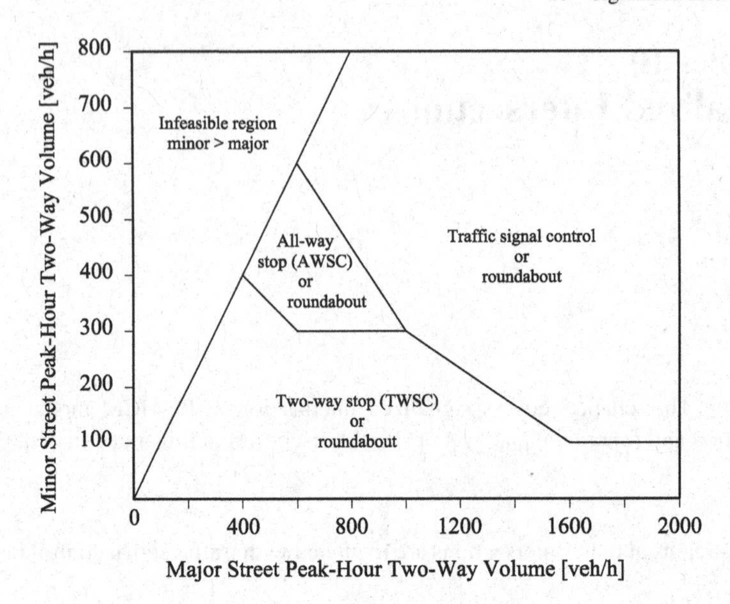

Fig. 10.1 Criterion for choosing the intersection control type in function of two-way peak-hour volumes on the major and minor streets [1]

- I (s) is *the intergreen time* during which vehicles must clear the intersection area and complete their manoeuvres before the next signal phase. It represents the time interval between the end of the green signal at one approach and the beginning of the green "go-ahead" signal at another. It results from the sum of the yellow (Y) time and the "all red" (AR) time in which all approaches have a red indication:

$$I = Y + AR \tag{10.2}$$

The intergreen can be calculated with the relation:

$$I = t_{pr} + \frac{v}{2 \cdot d} + \frac{L_1 + L_2}{v_1} \tag{10.3}$$

where:

- t_{pr} user perception-reaction time ($t_{pr} = 1.0$–1.2 s);
- v operating speed at the intersection (m/s);
- d vehicle deceleration ($d = 2.0$–$2.5\ m/s^2$);
- L_1 maximum trajectory length to be travelled to clear the intersection area;
- L_2 average vehicle length ($L_2 = 5$–6 m);
- v_1 mean speed along the trajectory L_1: 5–10 m/s.

In general, a yellow time $Y = 4$ s is set. Thus, the intergreen I is calculated with (10.3) and the all red time with (10.2) $AR = I - Y$.

Should the all red time $AR = 0$, by considering (10.2), from (10.1) it results that: $C = G + Y + R$;

R (s) is *the red time* during which the traffic lights are set to red for a certain movement or lane group.

Alternatively, the cycle length (C) can be calculated as:

$$C = ER + EG \tag{10.4}$$

where:

- ER (s) is the *effective red time:* time during which a traffic movement is not effectively utilizing the intersection [3]. The effective red time for a given movement or phase is calculated as $ER = R + t_L$, in which R is the displayed red time for a traffic movement in seconds and t_L is the total lost time for a movement during a cycle in seconds (see Fig. 10.2).
  The total lost time is calculated as $t_L = t_{sl} + t_{cl}$, where t_{sl} is the start-up lost time and t_{cl} is the clearance lost time;
- EG (s) is the *effective green time*: time during which a traffic movement is effectively utilizing the intersection.

It follows:

$$EG = G + Y - t_L \tag{10.5}$$

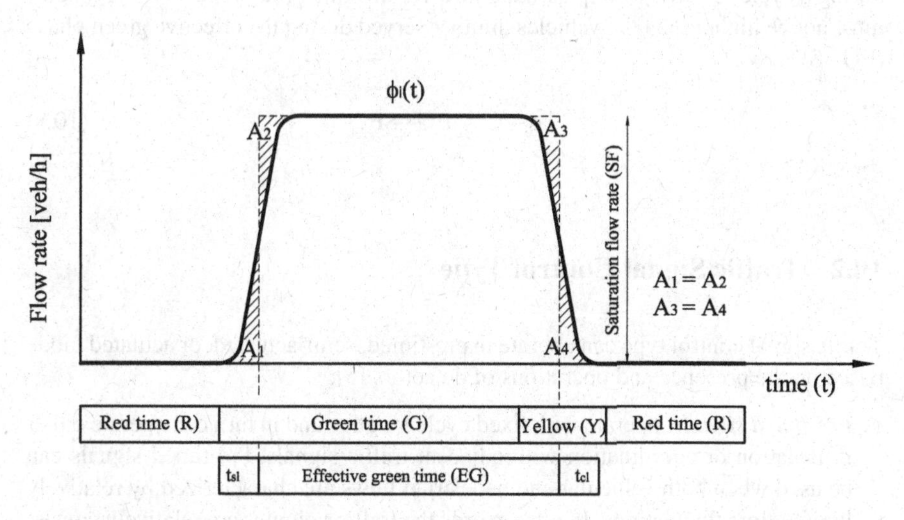

Fig. 10.2 Traffic flow variation at an entry of a signalized intersection

For a better understanding of the traffic signal phases described above, see Fig. 10.2. The flow of a generic traffic movement crossing the intersection can be represented with a function $\varphi_i(t)$ which starts from the green time, grows up to the saturation flow, keeps constant and then decreases during the yellow time up to zero before the red time starts again [4].

The number of vehicles crossing the intersection in the green and yellow times is equal to N_v:

$$N_v = \int_{[0,G+Y]} \phi_i(t) \cdot dt \tag{10.6}$$

The effective green EG is the green time which allows N_v vehicles to pass in the case of a sudden flow variation between the zero value and the *saturation flow rate* SF.

The *saturation flow rate* SF is the flow that can be accommodated by the lane group assuming that the green phase were displayed 100% of the time (i.e. EG/C = 1) [1].

Figure 10.2 clearly illustrates that the effective green is equal to:

$$EG = \frac{N_v}{SF} \tag{10.7}$$

In order for a vehicle movement not to be oversaturated (in other words, it is under-saturated or, at the most, in saturation condition), the arriving number of vehicles during a cycle ($q_i \cdot C$), with q_i demand flow for the lane group i and C cycle length) must not be higher than the vehicles number served during the effective green phase ($EG \cdot SF$):

$$q_i \cdot C \leq EG \cdot SF \tag{10.8}$$

10.2 Traffic Signal Control Type

Traffic signal control type can operate in pre-timed, semi-actuated, or actuated mode based on the presence and operations of detectors [5]:

- *Pre-timed signals*: operate with fixed cycle lengths, and in turn can operate either in isolation or coordination with adjacent traffic signals. Pre-timed signals can be used when both vehicular and pedestrian flows are characterized by relatively limited time fluctuations. In other words the traffic volumes are relatively consistent on a daily or day-of-week basis [5]. The cycle length can be modified each time but remains constant after every change;

- *Fully-actuated and semi-actuated signals:* the green time, allocated to each phase and the cycle length are completely influenced by the traffic demand evaluated by detectors.

An intersection that is under *fully-actuated* control has detectors on each arm. It is usually used at intersections of two major streets and where considerable variations exist in all approach traffic demand over the course of a day [3].

In case of *semi-actuated control*, detectors are usually placed on the minor street and on the major street left turns, but not on the major street through movements [5]. Green time is allocated to the major street until vehicles are detected on the minor street; then the green indication is briefly allocated to the minor street and then returned to the major street [3].

10.3 Phasing at Signalized Intersection

Phasing allows conflicting manoeuvres to be postponed and separated. In order to maximise the intersection capacity, a minimum number of phases is to be identified to guarantee safety. It requires that all the compatible traffic streams with similar green time lengths are unified.

The *stage control* identifies the sequential steps in which every manoeuvre is performed.

Figures 10.3 and 10.4 illustrate typical stages within two or three-phase systems for four-arm intersections [6].

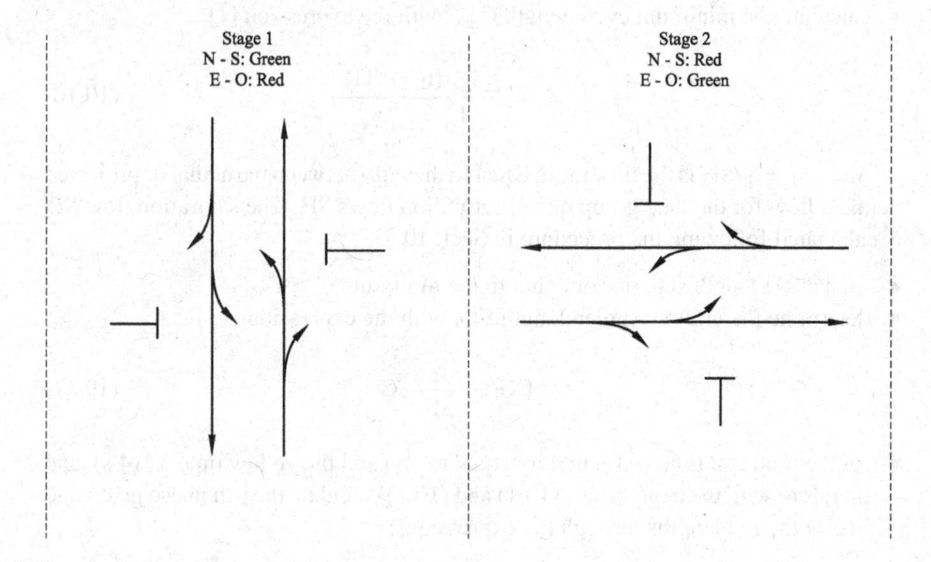

Fig. 10.3 Typical stages within two-phase system [6]

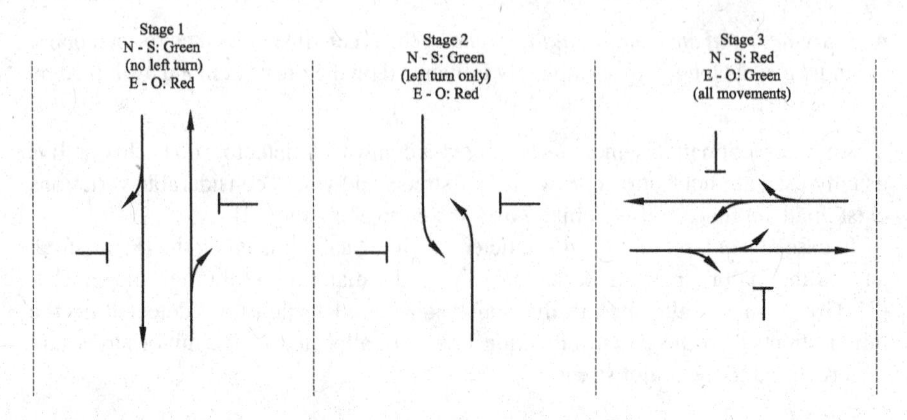

Fig. 10.4 Typical stages within three-phase system [6]

Should the green (G_i), yellow (Y_i) and all red (AR_i) times be known in each "i" of the "n" phases within the cycle, the corresponding cycle length (C) is computed with the following expression [1, 2]:

$$C = \sum_{i=1}^{n} (G_i + Y_i + AR_i) \tag{10.9}$$

On the other hand, in order to plan phases, it is necessary to proceed as follows:

- set the phase number on the basis of the intersection characteristics;
- calculate the minimum cycle length C_{min} with the expression [1]:

$$C_{min} = \frac{\sum_{i=1}^{n} (t_{Li} + TR_i)}{1 - \sum_{1}^{n} \gamma_i} \tag{10.10}$$

Since $\gamma_i = q_i/SF_i$ is the flow ratio, equal to the ratio between the actual or projected demand flow for the lane group q_i and saturation flows SF_i. The saturation flow SF_i is calculated following the procedure in Sect. 10.5:

- set a cycle length superior or equal to the minimum $C \geq C_{min}$;
- determine the effective green length EG_i with the expression:

$$EG_i = \frac{q_i}{SF_i} \cdot C \tag{10.11}$$

- set the total lost time t_{Li} (generally equal to 4 s) and the yellow time Y_i (4 s), and therefore with the expressions (10.1) and (10.2) calculate the i-th phase green and finally the red lengths through the expressions:

$$R_i = C - (G_i + Y_i).$$

Indicatively, the following values can be taken into account: $G_i \geq 15$ s for the main streams (for the others $G_i \geq 10$ s, for *fully-actuated and semi-actuated signals* $G_i \geq 5$ s); $Y_i = 3$–5 s (generally 4 s); $t_{Li} = 2$–4 s. The cycle must have a minimum length of at least 30 s ($C \geq 30$ s) and maximum length of 90–100 s for 2-phase cycles and 120 s for those with a higher number of phases.

10.4 Pedestrian Phases

Exclusive pedestrian phases or pedestrian and vehicle phases together can be scheduled for intersections in urban areas.

In the former case, the minimum green time can be calculated with [1]:

$$G_p = 3.2 + 0.27 \cdot N_p + \frac{L}{v_p} \quad \text{if } b \leq 3.00 \text{ m} \tag{10.12}$$

$$G_p = 3.2 + 0.81 \cdot \frac{N_p}{b} + \frac{L}{v_p} \quad \text{if } b < 3.00 \text{ m} \tag{10.13}$$

being:

G_p pedestrian green (s);
3.2 pedestrian start-up time (s);
N_p number of pedestrians crossing during an interval (ped);
L crosswalk length (m);
v_p the average speed of pedestrians (m/s) ($v_p \approx 1.2$ m/s);
b effective crosswalk width (m).

As for the simultaneous pedestrian and vehicle phases, it needs verifying that the sum of the green and intergreen times of the vehicular flow is sufficient for the pedestrian flow to pass at the same time.

10.5 Capacity Calculation

In order to obtain the average delay at the intersection of the lane group "i", the first step is to calculate the effective green ratio GR_i, as equal to the ratio between the effective green EG_i and the cycle C:

$$GR_i = \frac{EG_i}{C} \tag{10.14}$$

The capacity value of the lane group on which the same manoeuvre takes place is, according to the HCM manual [1] equal to:

$$c_i = SF_i \cdot GR_i \tag{10.15}$$

where FS_i denotes the saturation flow rate.

The saturation flow rate for subject lane group, expressed as a total for all lanes in lane group, can be calculated with the relation [1, 2]:

$$SF_i = SF_0 \cdot N \cdot f_b \cdot f_{tp} \cdot f_i \cdot f_p \cdot f_B \cdot f_a \cdot f_u \cdot f_D \cdot f_s \cdot f_{PD} \cdot f_{PS} \tag{10.16}$$

With $SF_0 = 1900$ veh/h/lane denoting the base saturation flow rate per lane.

The base conditions are: 3.60 m-long lanes, no heavy vehicles, horizontal approach, traffic equally subdivided between lanes of the same group, no parking spaces and bus stops within 75 m of the stop line, no right- or left-turns and no pedestrian flows.

Base saturation flow rate SF_0 is adjusted for a variety of conditions. The adjustment factors in (10.16) have to be computed with the expressions in Table 10.1 [1]. In order to consider any interference with pedestrian flows, a further multiplication coefficient k is used (see Table 10.1).

Once the saturation flows for lane groups have been obtained, the capacity can be calculated through (10.15) and consequently the saturation degree ρ_i corresponds to:

$$\rho_i = \frac{q_i}{c_i} = \frac{q_i \cdot C}{SF_i \cdot EG_i} = \frac{\gamma_i}{GR_i} \tag{10.17}$$

10.5.1 *Criterion for Calculating the Delay-Minimising Cycle Time*

During a signalized phase one or more lane groups have green light. A given flow ratio value $\gamma_i = q_i/FS_i$ corresponds to each lane group. The term *critical lane group* indicates the group with the maximum flow ratio value $\gamma_{c,i}$. Thus, the critical saturation ratio ρ_c can be defined as follows:

$$\rho_c = \frac{C}{C - (t_L + AR)} \cdot \sum \gamma_{c,i} \tag{10.18}$$

Since t_L denotes the sum of total lost time for critical movements and AR identifies the sum of the relative times of all red.

Once ρ_c value is set with (10.18), the cycle length is obtained with:

$$C = \frac{(t_L + AR) \cdot \rho_c}{\rho_c - \sum \gamma_{c,i}} = \frac{(t_L + AR) \cdot \rho_c}{\rho_c - \sum \left(\frac{q_i}{SF_i}\right)_c} \tag{10.19}$$

Table 10.1 Adjustment factors for saturation flow [1]

Factor	Formula	Definition of variables	Notes
Lane width	$f_b = 1 + \frac{b-3.60}{9}$	b = lane width (m)	$b \geq 2.40$ m; if $b > 4.80$ m a two-lane analysis may be considered
Heavy vehicles	$f_{tp} = \frac{100}{100+P_T \cdot (E_T - 1)}$	P_T = % heavy vehicles for lane group volume	$E_T = 2$
Grade	$f_i = 1 - \frac{i\,(\%)}{200}$	i = grade on a lane group approach	$-6\% \leq i \leq 10\%$ Negative is downhill
Parking	$f_p = \frac{N - 0.1 - \frac{18 \cdot N_m}{3600}}{N}$	N = number of lanes in lane group; N_m = number of parking maneuvers/h	$0 \leq N_m \leq 180$ $f_p \geq 0.050$ $f_p = 1$ for no parking
Bus blockage	$f_B = \frac{N - \frac{14.4 \cdot N_B}{3600}}{N}$	N = number of lanes in lane group; N_B = number of buses stopping/h	$0 \leq N_B \leq 250$ $f_B \geq 0.05$
Type of area	$f_a = 0.9$ in CBD $f_a = 1$ in all other areas	CBD = central business district	
Lane utilisation	$f_u = \frac{Q_g}{N \cdot Q_{g1}}$	Q_g = unadjusted demand flow rate for the lane group (veh/h) Q_{gl} = unadjusted demand flow rate on the single lane in the lane group with the highest volume; N = number of lanes in the lane group	
Right turns	$f_D = 0.85$ for exclusive lane $f_D = 1 - 0.15 \cdot P_D$ for shared lane $f_D = 1 - 0.135 \cdot P_D$ for single lane	P_D = proportion of right turns in lane group	$f_D \geq 0.05$
Left turns	Protected phasing: $f_s = 0.95$ for exclusive lane; $f_s = \frac{1}{1+0.05 \cdot P_a}$ for shared lane	P_s = proportion of left turns in lane group	For nonprotected phasing alternatives: specific calculation procedure [1]

(continued)

Table 10.1 (continued)

Factor	Formula	Definition of variables	Notes
Pedestrian-bicycle blockage	f_{PD}; f_P $k = \begin{cases} 0.95 \text{ if } q_p \leq 100 \text{ ped/h} \\ 0.85 \text{ if } q_p \leq 300 \text{ ped/h} \\ 0.75 \text{ if } q_p \leq 500 \text{ ped/h} \end{cases}$	q_p = pedestrian flow	For pedestrian and cycle flows with no phase: specific calculation procedure [1]

For fixed-cycle traffic lights, the delay minimization is obtained for $\rho_c \approx 0.8$–0.9.

Once the cycle length is known, the effective green can be calculated through (10.17):

$$EG_i = \frac{q_i \cdot C}{SF_i \cdot \rho_i} \tag{10.20}$$

10.6 Delay Calculation

The average control delay, experienced by all vehicles that arrive in the analysis period, for a given lane group can be calculated with the expression below [1]:

$$w_i = w_1 \cdot PF + w_2 + w_3 \tag{10.21}$$

being:

- w_i control delay per vehicle of movement "i" (s/veh);
- w_1 uniform control delay assuming uniform arrivals (s/veh);
- PF uniform delay progression adjustment factor, which accounts for effects of signal progression (Pre-timed signals PF = 1; actuated signals PF = 0.80–0.95);
- w_2 incremental delay to account for effect of random arrivals and oversaturation queues ($\rho_i > 1$), adjusted for duration of analysis period and type of signal control; this delay component assumes that there is no initial queue for lane group at start of analysis period (s/veh);
- w_3 initial queue delay, which accounts for delay to all vehicles in analysis period due to initial queue at start of analysis period (s/veh) (generally $w_3 = 0$).

Relation (10.22) gives an estimate of delay assuming pre-timed signals (PF = 1), uniform arrivals, stable flow, and no initial queue [1]:

$$w_1 = \frac{0.5 \cdot C \cdot (1 - GR)^2}{1 - [\min(1, \rho) \cdot GR]} \tag{10.22}$$

Relation (10.23) is applied to evaluate the incremental delay w_2 due to nonuniform arrivals as well as delay caused by sustained periods of oversaturation (oversaturation delay, see Chap. 8) [1]:

$$w_2 = 900 \cdot T \cdot \left[(\rho - 1) + \sqrt{(\rho - 1)^2 + \frac{4 \cdot \rho}{c \cdot T}} \right] \tag{10.23}$$

In which T is the analysis period expressed in hours. The reference period is usually 15 min; thus, T = 0.25.

Finally, the average delay for every approach w_a, as well as the average control delay for the intersection w_{int} can be computed:

$$w_a = \frac{\sum w_i \cdot q_i}{q_i} \tag{10.24}$$

$$w_{int} = \frac{\sum w_a \cdot q_a}{q_a} \tag{10.25}$$

In which:

- w_a delay for approach a (s/veh);
- q_i demand flow for the lane group i (veh/h);
- w_i delay for lane group i (on approach a) (s/veh);
- w_{int} delay per vehicle for intersection (s/veh);
- q_a demand flow for the approach a (veh/h).

10.7 Determination of Service Levels for Signalized Intersections

The HCM manual [1] identifies six levels of service (LOS) correlated with the average control delay per vehicle (Table 10.2) estimated for each lane group (w_i) and aggregated for each approach (w_a) and for the intersection as a whole (w_{int}); more precisely:

- LOS A: optimum traffic conditions. Most users cross the intersection without stopping;
- LOS B: good traffic conditions. Some vehicles are obliged to stop;
- LOS C: a non-negligible part of users stops at the intersection;
- LOS D: less favourable traffic conditions than the previous ones with a significant number of queued users waiting for crossing the intersection;
- LOS E: poor traffic conditions and high delays;
- LOS F: very poor traffic conditions, very high delays.

Table 10.2 Levels of service for signalized intersections

Control delay per vehicle (s/veh)	Level of service (LOS)
≤10	A
>10–20	B
>20–35	C
>35–55	D
>55–80	E
>80	F

10.7.1 *Case Study: LOS of a Signalized T-Intersection*

Consider a T-intersection outside the central business district area, with 3.50 m-wide lanes. For the generic movement "i" Fig. 10.5 shows the demand flow rate q_i, the heavy vehicle percentages P_{Ti}, the grade "i", together with the characteristics of the signal phasing. Determine the level of service (LOS) associated to every movement and LOS of intersection as a whole.

In order to solve the problem, the first and foremost step is to calculate the cycle length (C). By using the expression (10.9), there follows:

$$C = (28 + 4 + 3) + (24 + 4 + 3) + (24 + 4 + 3) = 97\,\text{s}$$

Consider a total lost time $t_L = 4$ s: through the expression (10.5) the effective green times can be calculated:

$$EG_1 = 28 + 4 - 4 = 28\,\text{s}$$

$$EG_2 = 24 + 4 - 4 = 24\,\text{s}$$

$$EG_3 = 24 + 4 - 4 = 24\,\text{s}$$

The green ratio times are obtained by the expression (10.14):

$$GR_1 = 28/97 = 0.289$$

$$GR_2 = 24/97 = 0.247$$

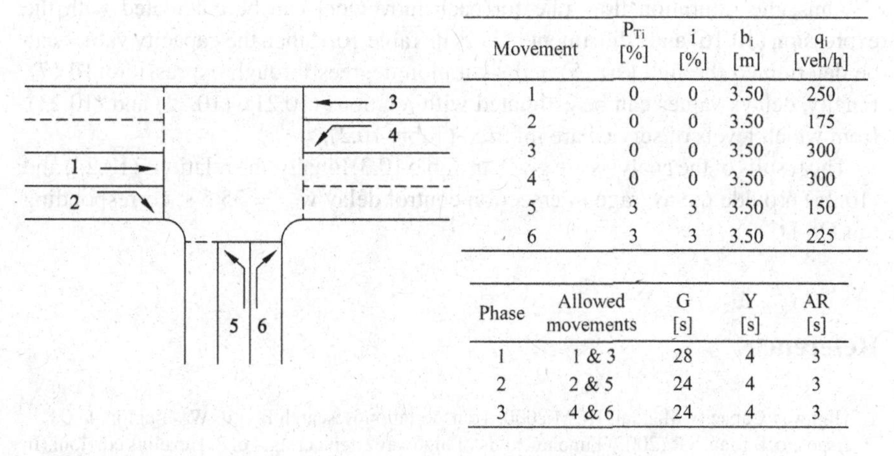

Movement	P_{Ti} [%]	i [%]	b_i [m]	q_i [veh/h]
1	0	0	3.50	250
2	0	0	3.50	175
3	0	0	3.50	300
4	0	0	3.50	300
5	3	3	3.50	150
6	3	3	3.50	225

Phase	Allowed movements	G [s]	Y [s]	AR [s]
1	1 & 3	28	4	3
2	2 & 5	24	4	3
3	4 & 6	24	4	3

Fig. 10.5 Intersection layout, geometric characteristics and signal phasing

Table 10.3 Results of the functional analysis

Movement	1	2	3	4	5	6
SF_0 (veh/h)	1900	1900	1900	1900	1900	1900
f_b	0.99	0.99	0.99	0.99	0.99	0.99
f_{tp}	1.00	1.00	1.00	1.00	0.97	0.97
f_i	1.00	1.00	1.00	1.00	0.99	0.99
f_p	1.00	1.00	1.00	1.00	1.00	1.00
f_B	1.00	1.00	1.00	1.00	1.00	1.00
f_a	1.00	1.00	1.00	1.00	1.00	1.00
f_u	1.00	1.00	1.00	1.00	1.00	1.00
f_D	1.00	0.85	1.00	1.00	1.00	0.85
f_s	1.00	1.00	1.00	0.95	0.95	1.00
f_{dp}	1.00	1.00	1.00	1.00	1.00	1.00
f_{ps}	1.00	1.00	1.00	1.00	1.00	1.00
SF (veh/h)	1878.89	1597.06	1878.89	1784.94	1706.96	1527.28
c (veh/h)	542.36	395.15	542.36	441.64	422.34	377.88
ρ	0.46	0.44	0.55	0.68	0.36	0.60
w_1 (s/veh)	28.31	30.85	29.20	33.02	30.12	32.22
w_2 (s/veh)	2.81	3.57	4.03	8.17	2.33	6.76
w_3 (s/veh)	0.00	0.00	0.00	0.00	0.00	0.00
w (s/veh)	31.11	34.42	33.23	41.19	32.44	38.97
LOS	C	C	C	D	C	D

$$GR_3 = 24/97 = 0.247$$

Thus, the saturation flow rate for each movement can be calculated with the expression (10.16) and adjustment factors in Table 10.1; then the capacity values can be determined through (10.15) and saturation degrees through expression (10.17). Finally, delays values can be estimated with relations (10.21), (10.22) and (10.23), from which levels of service are inferred (Table 10.2).

The results of the analysis are given in Table 10.3. Finally, the relations (10.24) and (10.25) provide the average intersection control delay $w_{int} = 35.5$ s, corresponding to LOS D.

References

1. Highway Capacity Manual: HCM 2000. Transportation Research Board. Washington. D.C.
2. Esposito T, Mauro R (2003) Fundamentals of highway engineering, vol 2. Hevelius edizioni (In Italian, available by info@hevelius.it)

3. Mannering FL, Washburn SS (2013) Principles of Highway Engineering and Traffic Analysis. Wiley, New York
4. Salter, R.J. Highway traffic analysis and design, Palgrave Macmillan UK, 1989
5. Signalized Intersections Informational Guide, 2nd edn. (2013). FHWA
6. Rogers M, Enright B (2017) Highway engineering, 3rd edn. Wiley, New York

Index

M. Guerrieri and R. Mauro, *A Concise Introduction to Traffic Engineering*,
Springer Tracts in Civil Engineering,
https://doi.org/10.1007/978-3-030-60723-4

Printed in the United States
by Baker & Taylor Publisher Services

Printed in the United States
by Baker & Taylor Publisher Services